D0455474

# PLANTS AND FLOWERS
# OF HAWAI'I

Published in North America by
University of Hawaii Press
2840 Kolowalu Street
Honolulu, Hawaii 96822

Simultaneously published in Singapore by
Les Editions du Pacifique
an imprint of Times Editions Pte Ltd
Times Centre, 1 New Industrial Road
Singapore 1953
© 1987 Times Editions Pte Ltd

Printed in Singapore

Library of Congress Cataloging-in-Publication Data
Sohmer, S.H.
        Plants and flowers of Hawaii.

        Bibliography: p. 159
        1. Botany — Hawaii.  I. Gustafson, R.
II. Title.
QK 473.H4S74 1986  581.9969  86-19231
ISBN 0-8248-1096-1

# PLANTS AND FLOWERS OF
# HAWAI'I

*S. H. Sohmer*
*and*
*R. Gustafson*

University of Hawaii Press
Honolulu

# Contents

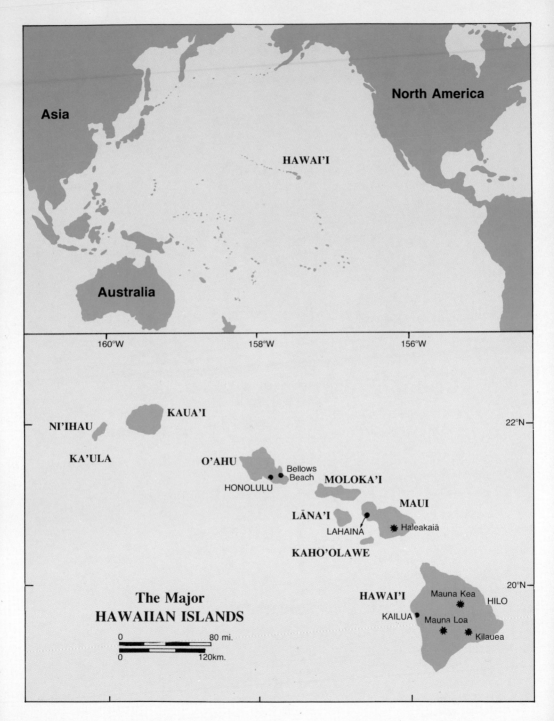

Asia

North America

HAWAI'I

Australia

160°W          158°W          156°W

NI'IHAU          KAUA'I                                                            22°N

KA'ULA

                    O'AHU
                              Bellows
                              Beach          MOLOKA'I
                    HONOLULU

                              LĀNA'I                    MAUI

                              LAHAINA          ✳ Haleakaiā

                    KAHO'OLAWE

The Major                    HAWAI'I          Mauna Kea                20°N
HAWAIIAN ISLANDS                                        ✳          HILO

                                        KAILUA ● Mauna Loa
          0          80 mi.                              ✳          ✳
                                                  Kilauea
          0          120km.

# INTRODUCTION

The Hawaiian archipelago is a chain of giant shield volcanoes that arose via the activity of a "hot-spot" below the ocean floor. This "hot-spot" has probably been actively erupting lava for perhaps as many as 70 million years, if the Emperor Seamounts are considered part of this chain. The Hawaiian Archipelago has formed as the ocean floor has moved across this "hot-spot" in a northwesterly direction. There are at present 132 islands, reefs and shoals with eight main islands. This is the largest assemblage of volcanic islands that are the most isolated from a continental landmass than any other similar set of islands in the entire world. An island arises if the flow of lava is sufficient over time to build it up above the surface of the sea. As the sea floor continues to move, the island is carried beyond the lava fountains from which it was formed. Wind and water wear away the landscape and the island presses down upon its supporting floor and subsides. The aging islands possess the rugged and spectacular scenic grandeur that looks so magnificent on picture postcards. What is left eventually sinks below the surface of the sea leaving in its wake what had been a fringing coral reef. This now becomes an atoll. Kure Atoll lies about 2,300 kilometers (about 1,500 miles) from the youngest of the Hawaiian islands, Hawai'i. Hawai'i rises more than 9,000 meters from the ocean floor to the peak of Mauna Kea with about half that height above the sea. The island of Hawai'i will eventually share the same fate as the older islands but now possesses a surface area of approximately 10,500 km$^2$. The oldest lava flow on the island of Hawai'i is said to be about 375,000 years before the present. The Midway Islands to the northwest are estimated to be about 27 million years old.

The islands offered to potential colonizers a wonderfully diverse topography, an equable climate and abundant rainfall and sunshine. The main islands lie approximately between 18° and 22° north of the equator. The trade winds from the north-

*Map showing the Hawaiian Islands and their setting in the Pacific. The main Hawaiian Islands lie between about 18°–22° north of the equator in the Pacific Ocean and comprise the southern end of the 2,300 km long Hawaiian Archipelago.*

*Page 4:* Argyroxiphium sandwicense *subsp.* macrocephalum. *Close-up of the flowers.*

east are forced to rise over the mountains, the windward slopes of which are generally wetter than the leeward slopes. Annual rainfall varies considerably depending on locality. For example, there is an average of about 60 cm (24 inches) rainfall per year in the city of Honolulu whereas about 400 cm (160 inches) fall at the head of Manoa Valley in the Ko'olau Mountains less than five miles away from the center of Honolulu. Temperature varies from an average annual range of about 17°C to 32°C (62°F to 89°F) at Waikīkī Beach to an average annual range of −8°C to 10°C (18°F to 50°F) on or near the summit of Mauna Kea (Armstrong 1973). What is believed to be one of the wettest places on earth exists in Hawai'i — the summit of Mt. Wai'ale'ale on the island of Kaua'i may receive as much as 15 m (49 feet) of rainfall in a given year.

## CAPTAIN COOK

On the 18th of January 1778 Europeans got their first glimpse of the Hawaiian Islands. The crew of the *Resolution*, Captain James Cook commanding, spotted the island of O'ahu that day. This was after a long voyage from Tahiti on their way to seek the "northwest passage" across the continent of North America. Because O'ahu lay to the windward of the *Resolution*, the vessel could not easily tack to it. However, the island of Kaua'i, the oldest of the principal inhabited islands of the Hawaiian chain, was seen shortly afterwards to the northwest. Nightfall found the *Resolution* about 30 miles off that island to the southeast. Men were soon to meet whose ancestors had gone

in separate directions tens of thousands of years earlier. The human dimension of this encounter was to be only part of the story, for the impact of contact would have a profound effect on the natural history of these islands.

The *Resolution* drew nearer Kaua'i on the 19th, at a place called Kīpū Kai. As that day went on, and the features of the island became clearer to the crew, natives in canoes came out to investigate this phenomenon. The expedition's artist, Mr. William Ellis, sketched the coast as the vessel drew closer. His watercolors became the first recorded views by Occidental man of the Hawaiian Islands — they are now at the Bishop Museum in Honolulu. Failing to see a suitable anchorage, Captain Cook kept about 1½ to 2 miles from the shore and proceeded up the leeward coast, where the ship found anchorage at Waimea Bay the following day after spending another night lying off the coast. The modern history of the Hawaiian Islands had begun.

We may assume from the watercolors produced by William Ellis and the accounts of naturalists accompanying subsequent expeditions that the vegetation was sparse on the leeward sides of the islands, i.e., the sides in the lee of the trade winds that blow nearly continuously from the northeast. There were reasons for the sparse vegetation on the leeward slopes that have become clearer within the past 20 or so years. The island discovered by the crew of the *Resolution* over 200 years ago had, of course, been discovered by man before. The numerous canoes filled with people that came out to see the

Europeans were proof of that. The population of the Hawaiian Islands, estimated at between 300,000 to 500,000 at the time of contact, originated from seafaring immigrants from the Marquesas and Tahiti about 1,000 to 1,500 years ago. The original immigrants could not have numbered more than several hundred. But these early immigrants found a unique and varied flora, far more so than that observed at European contact. The Polynesian success at populating the Hawaiian Islands had far more profound effects on the native flora and fauna than was originally thought.

What existed at the time the *Resolution* stood off Kīpū Kai on the island of Kaua'i was a native flora probably consisting of from 1,200 to 1,300 species of flowering plants. About 90% of these were not to be found anywhere else on the surface of the earth. It was a unique flora that was still filled with many wonderful examples of adaptation. Much of the leeward slopes were bare, however, and covered principally with *pili* grass (*Heteropogon contortus*), the grass used by the pre-contact Hawaiians to thatch the roofs and sides of their homes. *Pili* grass flourished on these leeward coasts and slopes probably because of man-made fire. There were no plant-eating

*Volcanism in Hawai'i. The "hot spot" below the ocean floor that gives rise to the lava fountains that have built the Hawaiian Archipelago has probably been active for at least 70 million years. Lava flows such as those shown here are collectively the mother of land in Hawai'i.*

terrestrial mammals in pre-contact Hawai'i. The flora at the time of contact included about 30 species of plants now believed to have been introduced by the Polynesians. These plants, where they flourished, had displaced whatever native plants had existed there. The Polynesian immigrants also brought the dog, a jungle fowl, the pig and the rat. The last was an uninvited guest. The pig and the rat have had a very significant and deleterious effect on the native flora and fauna. Pigs damage fragile ecosystems by rooting and allow introduced weeds to obtain strong footholds. The rat probably had its most significant effect on native birds and no doubt fed on the seeds of plants that were not used to such

*Captain Cook (1728–1779). The voyages of the famous Captain James Cook were wonders of discovery and models of decorum due to the profound leadership qualities of this man. There were earnest attempts made to learn about the people, places and biota met with during the voyages. Captain Cook came upon the Hawaiian Islands in 1778 while looking for the "northwest passage." He called them the "Sandwich Islands." Cook was treated as a god but a tragic misunderstanding resulted in a skirmish in which he was killed at Kealakekua Bay.*

*Artistic reconstruction of what a pre-Polynesian forest floor might have looked like with two now extinct flightless birds, one of which is an ibis, that are known through fossil evidence. Painting by Douglas Pratt.*

predation. Even the jungle fowl probably had a deleterious effect on native terrestrial birds. However, the single greatest ecological result of the successful Polynesian settlement was the elimination of most of the dryland and mixed mesic forests that covered much of the leeward sides of the islands. How these forests looked and what their full extent was we shall never really know, but they and most of the creatures that lived therein probably became extinct long before the arrival of Cook. We now know from fossil discoveries of nearly 50 extinct species of flightless birds that existed in Hawai'i (Olson & James 1984) and, as more fossil bones are found, we shall probably see that number rise. The Auwahi forest on the leeward slope of Haleakalā on Maui is probably one of the best remaining remnants of one aspect of such forests.

## A HUGE NUMBER OF ALIEN PLANTS

What has followed since contact have been assaults on the wet forests — forests of the mountains and windward slopes. These assaults have generally been less dramatic than the demise of the dry forests probably was but of potentially greater lasting effect. The huge numbers of alien plants and animals introduced since contact — many of which have become serious pests — have had a profound effect on the native flora. Alien insect pests and mammals chomp and chew, and bite and bore native plants in ways in which those plants have never learned to cope, having evolved without the influence of such creatures. Each airplane that

lands at Honolulu International Airport is a potential source of new pests to invade what is left of those ecosystems.

Introduced plants often do a lot better than native ones in Hawaiian habitats, particularly after such habitats have been disturbed. A native flora of between 1,000 and 1,100 species probably exists today with about 30% of them considered endangered or threatened and with several of them known only from a few surviving individuals. At least 10% of the native species present at the time the *Resolution* entered Hawaiian waters is now known or thought to be extinct. Others will no doubt unfortunately follow. The number of introduced plants that have adapted quite readily to Hawai'i's wonderful climate and topography over the past two centuries and become what we call "naturalized" has grown steadily. We mentioned that about 30 were present at the time of European contact. Approximately 150 were reported about a century later and over 800 are known today. The latter figure does not refer to all introduced species, which number well into the thousands, but only to those believed to be fully naturalized. We present in this book a summary of the native and naturalized flora of the Hawaiian Islands. A technical and complete revision of this flora is presently being concluded at the Bishop Museum.

*An* **Oreobolus** *bog in the Upper Hana Forest Reserve. This photograph was taken in 1973 and shows well-formed tussocks of this sedge.*

*The same bog photographed from near the same position in 1981. Pigs have nearly destroyed the bog.*

# A SHORT HISTORY
# OF BOTANY IN HAWAI'I

The first botanists of Hawai'i were the Polynesian immigrants. They brought a knowledge of the plants of the areas from which they came and over the centuries that followed, as their descendants became what were eventually to be called "Hawaiians," they developed a knowledge of native plants. Many of those plants were similar to those they had known. There were few native plants at the time of Polynesian settlement that could have been used as food, but as time went on the medicinal and other value of many plants became known through trial and error.

The native Hawaiians organized their knowledge of the plant life of Hawai'i through an oral tradition that was basically preserved in the knowledge of individuals called *Kahuna la'au lapa'au* — priests who possessed knowledge of the medicinal and other uses of plants. Nearly all individuals had some knowledge of plants, and nearly all indigenous plants had specific names.

There is now general agreement that 27 plant species were

*The noni (Morinda citrifolia) provided the Hawaiians with a variety of uses including a food source in times of scarcity.*

introduced during the prehistoric Polynesian immigration period. Six others are not universally agreed to by ethnobotanists. Table 1 provides information for 26 of these based largely on the work of Dr. Isabella Abbott, an ethnobotanist at the University of Hawaii. Scientific, Hawaiian and English common names are provided as well as the principal uses of the plants and the plant part(s) used. The reason some disagreement exists is that some of the introduced plants may have become so well adapted to the Hawaiian landscape that it is difficult or impossible to distinguish them as non-native. Several weeds, for example, may have been introduced inadvertently at this time.

Twelve of the plants listed in the table are food plants, not counting sugar cane and 'awa, and two of these, *kalo*, most commonly called *taro* in Hawai'i, and *niu* (coconut), were of prime importance. *Taro* was known and grown by all Polynesian peoples, save the *Maori* who either could not grow it effectively in much of New Zealand's climate or who did not take propagules with them on their voyages of discovery. The Hawaiians used *taro* extensively in the form of *poi*. *Poi* is made when *taro* corm is mashed to a pulp, thinned with water, and strained. As *poi* it was the principal source of starch for pre-contact Hawaiians filling much the same niche as rice, wheat, corn, rye, oats and potatoes fill in other cultures. *Taro* is usually cultivated in small, irrigated plots under conditions much like those necessary for the cultivation of rice.

*Kapa*, the Hawaiian version of bark cloth, was made by beating the specially prepared bark of *wauke*.

## BOTANY AFTER COOK'S VOYAGE

Western science also came to Hawai'i with Captain Cook. Botanical science had been codified and sanctified by many centuries of development in Europe. The discovery of the Pacific Islands one right after another, culminating with the discovery of the Hawaiian Archipelago opened yet another cornucopia for European botanists. The Hawaiian flora would be collected and studied for well over a hundred years before a decent understanding of the plants would emerge. It is modern botanical science that has permitted knowledge of the plant life of Hawai'i to be transferred on a worldwide basis. Paradoxically, and tragically, it is the coming of the outside world that has sounded a death knell to many of Hawai'i's unique flowering plants.

The first collections of plants from Hawai'i were made during Cook's voyage, his third, which touched the islands of Kaua'i, Ni'ihau and Hawai'i.

Specimens were collected by David Nelson during an attempt to reach the summit of Mauna Loa in 1779. The collections are in the British Museum with some duplicates at Bishop Museum. Dr. Harold St. John has published papers on the species represented in this collection after many hours spent searching for them at the British Museum. St. John considered 15 of these collections to represent new species that have presumably become extinct in the interim.

The voyage of the *H.M.S. Blonde* under Captain Lord Byron with James Macrae as botanist took place in 1825. Macrae was the first botanist to describe the Hawaiian silversword, probably the most famous of the endemic Hawaiian plants.

E. Alison Kay in *A Natural History of the Hawaiian Islands* (1972) has summarized the early exploring voyages and provided insight into what these exciting glimpses of an exotic land meant for northern Europeans. The botanical specimens collected on these voyages eventually found their way into major herbaria. Collections made by Berthold Seeman in 1847, for example, formed the basis of European knowledge about the flora of Fiji. The Russian voyages commanded by Otto von Kotzebue in 1815 and 1823 brought back many specimens from Hawai'i, especially those collected by Chamisso. Charles Gaudichaud-Beaupré made one of the most important early collections in Hawai'i with a French expedition aboard the *Uranie*. He described many new species and genera of plants in Hawai'i. The Wilkes Expedition sent by the United States spent about six months in Hawai'i in

**Table 1.** *Plants introduced to Hawai'i during the Polynesian Period (After Dr. Isabella Abbott)*

| Scientific Name | Hawaiian/English Names | Principal Uses | Plant Part Used |
|---|---|---|---|
| *Aleurites moluccana* | *kukui*, candlenut | fuel, lighting, wood, oil, medicine, dyes | fruit, stems, roots |
| *Alocasia macrorrhiza* | *'ape* | famine food | stem |
| *Artocarpus altilis* | *'ulu*, breadfruit | food, wood, dye, drums surfboards | fruit, wood, bark |
| *Broussonetia papyrifera* | *wauke*, paper mulberry | fiber for tapa, cordage | stem |
| *Calophyllum inophyllum* | *kamani* | wood, oil, perfume | stem, fruit, flowers |
| *Cocos nucifera* | *niu*, coconut | food & drink rope, drums | fruit, husk, stem |
| *Colocasia esculenta* | *kalo, taro* | food (poi, luau), medicine | stem, leaves |
| *Cordia subcordata* | *kou* | utensils wood, dye | stem, leaves |
| *Cordyline terminalis* | *ki, ti* | food preparation, wrappers, inside thatching | leaves |
| *Curcuma longa* | *'olena*, turmeric | dye, purification, medicine | stem |
| *Dioscorea alata* | *'uhi*, yam | food | root |
| *Dioscorea bulbifera* | *pi'oi*, yam | food | root |
| *Dioscorea pentaphylla* | *pi'ia or pi'a*, yam | food | root |
| *Syzygium malaccense* | *'ohi'a'ai*, mountain apple | food, wood, medicine | fruit, stem |
| *Hibiscus tiliaceus* | *hau* | cordage | stems |
| *Ipomoea batatas* | *'uala*, sweet potato | food, medicine | stem, leaves |
| *Lagenaria siceraria* | *'ipu*, gourd | containers, drums | fruit |
| *Morinda citrifolia* | *noni* | medicine, dyes, famine food | fruit, stem and root |
| *Musa acuminata* hybrids | *mai'a*, banana | food, cooking, cordage, inside thatching | fruit, leaves, sheath |
| *Piper methysticum* | *'awa* | drink, medicine | root |
| *Saccharum officinarum* | *ko*, sugar cane | sugar | stem |
| *Schizostachyum glaucifolium* | *'ohe* | lamps, water containers, musical instruments, tapa stamps | stem |
| *Tacca leontopetaloides* | *pia*, arrowroot | food (starch) | root |
| *Tephrosia purpurea* | *'ahuhu* | fish poison | whole plant or root |
| *Thespesia populnea* | *milo* | wood | stem |
| *Zingiber zerumbet* | *'awapuhi*, shampoo ginger | medicine, shampoo | fruit |

1840-41 where a team of scientists made significant collections. The famous botanist Asa Gray worked on the collections made by this expedition in Hawai'i. After the middle of the 19th century the exploring voyages essentially ceased and new collections came into the scientific world via the activities of local naturalists who were often the scions of the missionary families who began arriving in Hawai'i in the 1820's. These individuals made up in interest and long-term residency what they lacked in scientific training. Gradually, however, scientists came to the islands for long periods of time. The collections of Mann and Brigham, Hillebrand (most of his material to be eventually destroyed in Berlin during World War II), and Heller, to mention just a few, are important examples.

William Hillebrand deserves attention. He came to Hawai'i in 1851 after several years spent wandering from Germany to Australia, the Philippines and San Francisco in search of a climate more suitable to his health. In Hawai'i he served as physician at Queen's Hospital as well as private physician to King Kamehameha V. During the 20 years he spent in Hawai'i he devoted his free time to the study of its flora. He proceeded to gather one of the most important collections of Hawaiian plants and he relentlessly explored all the islands to this end. He also created a lush garden around his home in Honolulu, part of which is now the City and County of Honolulu's Foster Botanical Garden. He returned to his native Germany in

*William Hillebrand.*

1871 and worked on assembling his knowledge and the collections he had made during his residence in Hawai'i to write the first and, to this date, only flora of the Hawaiian Islands. He died in 1886 and his flora was published posthumously in 1888. To this day his work has served generations of students of the Hawaiian flora.

In this century the collections made by Charles Forbes, the first full-time botanist appointed at Bishop Museum in 1908, form the historical basis of the Hawaiian section of that museum. Dr. J. F. Rock is another individual of special interest to all students of Hawaiian plants. Austrian by birth, Rock came to Hawai'i in 1907 at the age of 23. He spent much of the years between 1908 and 1920 working with the Hawaiian flora and for 9 of those years was employed at the College of Hawaii. It was during this time that he made one of the best collections of Hawaiian plants that exists today, as well as wrote two of his most important treatises on Hawaiian plants: *The Indigenous Trees of the Hawaiian Islands* (1913) and *A Monographic Study of the Hawaiian Species of the Tribe Lobelioideae, Family Campanulaceae* (1919). In 1920 he left Hawai'i and began three decades of plant exploration in Asia, particularly China, where he became the world's expert on the Nakhi tribe in Western Yunnan. He collected thousands of plants for the U. S. Department of Agriculture during this time. He eventually returned to Hawai'i where he died in 1962.

As of this writing two of the longest-lived botanists that have lived and worked in the Hawaiian Islands are still active.

Dr. Otto Degener came to Hawai'i in 1922. In 1932 he began the series known as *Flora Hawaiiensis* which is now a collection of some seven books in which line drawings of nearly every species included have been provided along with keys and descriptions.

Dr. Harold St. John arrived in Hawai'i in 1929 to teach at the University of Hawaii. He retired from the university in 1959. Visitors to the University of Hawaii will find the Botany Department housed in St. John Hall. He has been pursuing botanical research full time at the Bishop Museum ever since. His student, Dr. F. R. Fosberg, has been another potent force in the study of Hawaiian flowering plants.

All of these individuals and many others have provided the collections without which our understanding of the Hawaiian flora today would not be possible. In Hawai'i this has particular poignancy, for, of the plants known as a result of the collections described above and studied in the historical period since the arrival of Captain Cook, as mentioned earlier, over 30% are now endangered or threatened and over 10% are known or presumed to be extinct.

Taro *fields in Waipi'o Valley on the Big Island. Fields like these were at one time common throughout the islands where there was enough water to nourish wet* taro *varieties. Waipi'o Valley supported a large population in pre-contact times. This population was protected by the sheer walls of the valley that extend to the sea. The inaccessibility of the valley eventually led to its decline as a center of human settlement after unification of the Islands under King Kamehameha I.*

# BIOLOGY AND DERIVATION OF THE HAWAIIAN FLORA

We now believe that there are about 1,000 native species of flowering plants in Hawai'i. There are 88 families and 211 genera represented by these species. The 16 largest of the genera account for nearly 50% of the native species. Over 90% of these species are endemic, the largest percentage of endemic species found in any flora. Nearly all of the native species, and certainly all the endemic species, evolved in the Hawaiian Archipelago. The interesting question is from whence did they or their ancestors come?

## DERIVATION OF THE HAWAIIAN FLORA

It is generally held that life came to these islands via wind, water, or wing. There was no other way for plant or animal dispersal to take place before the advent of man. Nowadays more plants come via 747 jumbo jet than any other means and they come via willful (and in most cases illegal) introduction or as stowaways — a seed caught on

*The ti plant (***Cordyline terminalis***; kī) is an old Polynesian introduction with many economic uses.*

the heel of a shoe, a burr stuck on the hem of a skirt, etc. These islands were barren lava rock when they first rose above the surface of the sea. If the Hawaiian Archipelago is, as mentioned, as much as 70 million years old (or at least 27 million years old) then plants have had at least that long to arrive. It is therefore not surprising that an event that has such a low probability of occurring — a seed arriving by being blown in thousands of miles by a storm, or attached to the body of a bird blown off its normal course, or, presently least likely, washed ashore after floating many weeks in seawater, etc. — will occur relatively frequently over that period of time. Indeed, it is hypothesized that such an event occurred nearly 300 times during the history of the archipelago to provide for the native flora we know today. An analysis of the flora of the Hawaiian Islands in 1948 led Dr. F. R. Fosberg to hypothesize that that flora had been derived over time via the establishment of 272 successful introductions. Dividing this number into the number of years available in which such natural introductions could have occurred if one assumed only the span

21

of time in which the present main islands have been in existence leads to the conclusion that, on average, one successful introduction need have occurred once every 20,000–30,000 years in order to account for the flora we see today. If one takes 70 million years as the life of the entire archipelago from Hawai'i Island to the Aleutian Trench, such an event need not have occurred more than once every 250,000 years or so, as Fosberg pointed out, based on the flora presently known. Of course, many species have become extinct over this time, and the total number of successful immigrants, were all species *ever* to have evolved in the Hawaiian Archipelago to be known, would doubtless be much higher than the number hypothesized on the basis of the presently known flora. It is interesting here to compare the available statistics for other groups of plants and also animals with the postulated figures for flowering plants, as we do in Table 2.

The biogeographic relationships of the flowering plants of Hawai'i are principally to the southwest. In other words, it seems that the largest number of successful colonists came from that area than from any other source area. This probably has to do with the jet stream and directional patterns of migratory birds. Although the actual percentages will probably change as our knowledge of the flowering plants is brought up to date by the botanists working on a new *Manual of the Flowering Plants of Hawai'i*, the overall significance of the contributions made to the

**Table 2.** *Numbers of presumed original colonists, derived native species, and endemic species for a selection of the Hawaiian biota. (Source of table, including notes, Wagner and Gagné.)*

| Animal or Plant Group | Estimated Number of Colonists | Estimated Number of Native Species | % Endemic Species |
|---|---|---|---|
| Marine algae[1] | ? | 420 | 13 |
| Ferns and fern allies[2] | 114 | 145 | 70 |
| Mosses[3] | 225 | 233 | 46 |
| Flowering plants | ca. 270[4] | ca. 1000[5] | 91 |
| Terrestrial mollusks[6] | 24–34 | ca. 1,000 | 99 |
| Marine mollusks[7] | ? | ca. 1,000 | 30–45 |
| Insects[8] | 230–255 | 5,000 | 99 |
| Mammals | 2 | 2 | 100 |
| Birds[9] | ca. 25 | ca. 135 | 81 |

[1] Chock, A. K. 1968. Hawaiian ethnobotanical studies I. Native food and beverage plants. Econ. Bot. 22: 221-238.

[2] C. H. Lamoureux (pers. comm.). Endemic species number includes four taxa previously considered species but now reduced to subspecies. The figure also includes one species of *Schizaea* only known from fossil spores.

[3] Hoe, W. 1979. The phytogeographical relationships of Hawaiian mosses. Ph.D. thesis, University of Hawaii.

[4] Fosberg, F. R. 1948. Derivation of the flora of the Hawaiian Islands. *In:* Zimmerman, E. C., Insects of Hawaii, Vol. 1, pp. 107-119.

[5] Wagner, W. L., D. R. Herbst, and S. H. Sohmer. Manual of the Flowering Plants of Hawai'i. In prep.

[6] C. C. Christensen (pers. comm.)

[7] E. A. Kay (pers. comm.)

[8] Gagné, W. C., and C. C. Christensen. 1985. Conservation Status of Native terrestrial invertebrates in Hawaii. *In:* Stone, C. P. & J. M. Scott, eds., Hawai'i's terrestrial ecosystems: Preservation and Management, Coop. Natl. Park Resources Stud. Unit, Univ. of Hawaii, Honolulu. pp. 105-126.

[9] Shallenberger, R. J., ed. 1984. Hawaii's birds. Hawaii Audubon Society, Honolulu. 96 pp. Note: Bird figures include species and subspecies. Also included are 25 species of seabirds that breed in Hawai'i.

Hawaiian flora by the different source areas will probably not change significantly from that picture presented by F. R. Fosberg. Of his hypothesized 272 original colonists he postulated that about 40% came from the Indo-Pacific, about 16–17% from the south (his Austral affinity), about 18% from the American continent, and less than 3% from the north. Fosberg considered about 12–13% to be either cosmopolitan or pantropical and not indicative of any particular area. He could not address himself to the biogeographical affinities of about 10% of the flora.

There are groups of organisms, like ferns and mosses, in which the estimated number of original colonists have not evolved a much larger number of descendants as is the case in flowering plants and insects. Those that are in the former group are also those that have the smallest percentage of endemic species. This probably has to do with the mode of dispersal; if the group produces small easily transportable propagules, the spores of ferns and mosses as examples, then it is unlikely that a successful colonist will remain isolated from its parental source. The descendants of the colonist, therefore, will probably not have the potential to change as readily or as rapidly as the descendants of those colonists that are genetically isolated from their parental source.

## A DISHARMONIC FLORA

Looking at the native flora in comparison with continental source areas we discover one of the basic facts concerning the flowering plants of Hawai'i. The native flora of Hawai'i is not fully representative of the floras of the nearest continental areas. Whole families and groups are missing. When this is analyzed, one comes to see that the groups missing are those in which the propagules — the seeds or fruit — are either too large or too clumsy to be moved by wind, water or wing. The groups not represented in the native flora of Hawai'i are those that are generally not easily dispersed. Therefore, the native flora of Hawai'i is representative of those groups that at present, or at least in the past, produced small, light, or otherwise movable seeds or fruit. Hawai'i's flora is therefore considered "disharmonic."

Sherwin Carlquist, more than any other recent biologist, has heightened our awareness of island biology, and has summarized the principles of plant dispersal in the Pacific in a number of papers and books (1970, 1974). For Hawai'i he hypothesized a number of original flowering plant immigrants and calculated the dispersal modes for these. The most significant mode of dispersal is postulated to be internal carriage by birds. The least significant mode is considered to be air flotation. Other individuals have considered that seawater flotation has been the least probable mode of dispersal to Hawai'i. They point out that the present pattern of ocean movements does not allow for currents to come to Hawai'i from any place with which Hawai'i shares a flora. However we do not necessarily know what the ocean currents might have been like in the geological past, when the continental configurations were different from what they are today.

**Table 3.** *Modes of long-distance dispersal of the flowering plant immigrants assumed to have given rise to the presently known native flora of the Hawaiian Islands (data from Carlquist 1974).*

| Dispersal Mode of Seeds/Fruits | % plants hypothesized derived in Hawai'i |
|---|---|
| Air flotation | 1.4% |
| Birds | |
|    mechanically attached | 12.8% |
|    eaten & carried internally | 38.9% |
|    embedded in mud on feet | 12.8% |
|    attached to feathers by viscid substance | 10.3% |
| Oceanic Drift | |
|    Frequent (able to float for prolonged periods) | 14.3% |
|    Rare (probably resistant to seawater but unable to float | |
|       for prolonged periods and likely to arrive by rafting) | 8.5% |

## FLORAL BIOLOGY

Getting here was only the first major hurdle that potential immigrants had to overcome. Once here they had to become established. A propagule could land in an unsuitable environment. If so, all the effort expended in getting here was wasted. The successful colonists, therefore, are only representative of those colonists who, literally, were able to set down roots. This fact of life would also tend to favor those easily dispersed plants that have the ability to germinate rapidly and establish themselves quickly. Many "weedy" species exhibit these appropriate characteristics and many of the present-day native plants of Hawai'i have probably been evolved from ancestors that we would have called "weeds."

If the colonist arrives and becomes established there is yet another potential problem to overcome, for the odds are less favorable that two such colonists would arrive and become established at the same time. Plants that require a pollen parent and seed parent in order to reproduce, therefore, would be much less likely to become established and to begin the process of evolution that would result in one of the groups of species we see today. Botanists use the term "dioecy" to indicate the breeding situation in which the pollen (which produces the male gamete or sperm) and the ovule (which produces the female gamete or egg and which ripens into what we call the seed) are produced on *separate* individual plants. In such a situation self-pollination and, therefore, self-fertilization is not possible. How unlikely it is that two such individuals of a species would have become established in Hawai'i at the same time after a long-distance dispersal event. There are many

*Male plant of* **Coprosma waimeae** *(ōlena). The genus* **Coprosma** *(pilo) is found throughout the Pacific. The Hawaiian members of the group are dioecious: individual plants produce only staminate (male) or pistillate (female) flowers. This shows a staminate plant with filaments, or stamen stalks, that are very long and hanging.*

flowering plants that produce both pollen (in stamens) and ovules (in pistils) on the same individual and such a situation is called "monoecy." Most monoecious plants have bisexual or perfect flowers, i.e., stamens and pistils produced in the same flower. Such individuals are often what we call self-incompatible, i.e., the pollen of that plant cannot function on that plant. If this is an absolute isolating mechanism in that species, the same thing said about the dioecious condition above applies here. Therefore, it is hypothesized that, over time, those plants that were monoecious and which were not absolutely self-incompatible were more likely to become established in the Hawaiian Islands, all other things being equal. A great deal of thought on such matters has already been expended by many great minds, most notably those of Alfred Russel Wallace and Charles Darwin.

# HYBRIDIZATION

There has not been much experimental work with elements of the Hawaiian flora, but what there has been has confirmed the impression based on morphological studies that Hawaiian plants seem to hybridize a lot. The species of many genera appear to be able to hybridize and produce intermediate offspring. However, ecological differences in environment, for which the parental types are best suited, allow the parental types to out-compete their hybrid progeny and this keeps the species morphologically distinct. Perhaps there are sound biological reasons why hybridization on relatively small

islands takes place so frequently. Perhaps we are wrong in assuming that there is so much of it. However, if our observations are correct and if we can extrapolate the experimental evidence we do have, we can safely assume that hybridization is a phenomenon well represented in the native Hawaiian flora. One hypothesis for this phenomenon is that the tendency to hybridization is important in a situation where the size of populations of species, and the number of such populations, will of necessity be relatively small because of the island setting. Hybridization under such conditions would tend to maximize the amount of variability available to that group and perhaps tend to thereby maximize a "resistance to extinction." Even in one of the groups of native Hawaiian plants that has most often hitherto been cited as an example of a situation where there are many small distinctive species with little or no hybridization occurring between them, *Cyrtandra*, we now find that hybridization may indeed be quite common. Many of the entities named as species in the past are probably examples of this hybridization — the variation simply having been misinterpreted.

One of the apparent paradoxes in the study of the native Hawaiian flora is that such a relatively high proportion of the species are dioecious·or approaching that condition. This can easily be observed by anyone. This goes against the hypothesis that

**Gunnera petaloidea** (*'ape 'ape*) *is a giant herb with colonies of blue-green nitrogen-fixing algae in its stems.*

monoecious, self-compatible species would have had the most likely chance of becoming established in the Hawaiian Islands during geologic time. We think that part of the explanation for this is that *after* establishment and in the process of adaptation and change over the course of thousands, hundreds of thousands, or millions of years, the conditions in these islands would have favored breeding systems that favor out-crossing over self-fertilization for the same reasons hybrids seem to be so commonly encountered.

There are many forms that outbreeding systems can take other than the ultimate one: having separate pollen and ovule individuals. We need not detail them here other than to note that in a species with perfect flowers one can have either of two conditions prevailing: (1) two types of individuals present — one having flowers with functional stamens and pistils, the other with either functional pistils only or functional stamens only; (2) situations where both stamens and pistils are functional but one or the other matures first so that self-fertilization remains impossible despite both parts being present and functional. The floral biology of Hawaiian flowering plants is still barely known despite study by a number of botanists.

## LOSS OF DISPERSIBILITY

Another remarkable feature of Hawaiian flowering plants is the relatively large size of the fruit and/or seeds produced by many of the native tree species. This is remarkable since one of the basic tenets of long-distance

dispersal is that the ancestors of these plants would most likely have had small, easily dispersed fruit or seeds. Obviously, there has been a loss of dispersibility in the Hawaiian flora. Once the original colonists became established an evolutionary premium was placed on *not* producing easily dispersed fruit or seeds. Hawai'i stands out in this regard more than any other island group in the Pacific, or the entire world for that matter. Again, Sherwin Carlquist has documented this fact in *Hawaii, A Natural History*. He has shown how in nearly all *Bidens* spp. (ko'oko'olau) the fruit has lost the long barbed awns that occur at the top of the fruit in nearly all continental species of this genus. These awns are what attach the fruit to the feathers or fur of birds and mammals, not to mention the clothes of present-day *Homo sapiens*. In addition the fruit of endemic Hawaiian members of this genus have become much larger and twisted for some reason. In any case the awns are gone or if they remain they are barbless. Those species of ko'oko'olau that grow in the wetter forests are also those that show the largest and least dispersible fruit. The same sort of syndrome can be found in nearly every group of Hawaiian plants when compared with members of that same group from other Pacific islands. The Hawaiian representatives will have the largest, least dispersible fruit and will tend to have lost whatever fea-

---

**Pritchardia** *palms* **(loulu)** *in tall 'Ōhi'a forest on Hawai'i.*

*Pēpē'ōpae Bog on Moloka'i. There are bogs on all the main islands, except Kaho'olawe Lāna'i and Ni'ihau.*

tures exist that aid in the "portability" of related species found elsewhere.

There are even clearly marked trends within groups that have a number of native Hawaiian species that are found in different habitats and islands. As a general rule those groups with species occupying habitats from seashore to mountain forests will tend to show the largest fruits produced by those living in wet forests. Those species of such a group occupying coastal habitats will tend to have the smaller, portable fruits. This tendency occurs even within the archipelago, for, in groups with species distributed throughout the present-day main islands, the largest-fruited species with the smallest number of "portability" features occur on the older islands of Kaua'i and O'ahu. This no doubt has to relate to the fact that the younger islands are themselves more recently colonized and therefore would tend to have a larger number of species that have retained "portability" factors than on the older islands of the archipelago.

One of the most often cited reasons for the large fruit and seeds of wet forest species is that only large seeds provide the room for ample stored food for the embryo and the young plant it develops into to grow on. Since the juvenile plant will find itself pretty much in the shade in such a habitat it will require sufficient support to get it to the point at which it will be able to modestly compete for some space and light. Plants of open environments do not generally have such factors to contend with.

Another factor that probably bears upon the loss of dis-

persibility as species evolve from open coastal habitats into closed wet forest ones is the loss of the dispersal agent. Species that are dispersed by particular agents (wind, water or wing) will, as they evolve into new habitats, often leave those agents behind. Evolution, as a general rule, is usually very efficient. When there is no longer a "need" for something it is more efficient not to produce that something and eventually it is lost.

## WOODINESS IN THE FLORA

As mentioned earlier, it is commonly held that the plants most likely to have been successful colonists were "weedy" plants. More of them were prob-

ably herbaceous, suffrutescent or shrubby than tree-like. Some families represented principally by herbaceous species in continental source areas are only represented by woody or near woody species in Hawai'i. Such families as the Amaranthaceae and Chenopodiaceae are good examples. The endemic species are only woody or at least suffrutescent.

Carlquist (1970) gave excellent examples in the genus *Chamaesyce (Euphorbia)* showing the progression from herb to tree with such species as *C. degeneri* (an herbaceous mat), *C. celastroides* and *C. multiformis* (shrubs) to *C. rockii* (tree).

There are at least several reasons for the preponderance of woodiness in the native Hawaiian flora. One of them might be the general tendency of many groups of plants to continue to grow in a climate such as Hawai'i's where there are no winters to bring things to a halt. Another might be, as mentioned above, that whereas "weedy" plants may have been favored as colonists, there would have been strong tendencies to evolve tree-like species that would be better able to compete for light and space in the forests, particularly the wet forests, of the evolving Hawaiian landscapes.

Opposite page: **Sadleria cyatheoides** (**'ama'u**) *is an endemic fern common on all of the islands.*
*The shrubby plant being studied below in Ma'akua Gulch on O'ahu is a member of the Amaranthaceae. It is a member of the genus* **Charpentiera,** *called* **pāpala** *in Hawaiian.*

# EVOLUTION OF SPECIES AND LANDSCAPES

## ADAPTIVE RADIATION

One of the best recorded examples of adaptive radiation in flowering plants is that of the California tarweeds and the Hawaiian *Dubautia-Argyroxiphium-Wilkesia* complex. This is a remarkable example of adaptation. There is little resemblance between the magnificent Hawaiian silverswords and a group of plants found on the coast of California commonly called tarweeds because of the sticky resin they produce. However, *Adenothamus validus*, from Baja California, is a species that may resemble the ancestors of both the California tarweeds and the three genera mentioned above.

The ancestral tarweed that is hypothesized to have landed in Hawai'i was most likely a shrub. Its descendants radiated into nearly all Hawaiian habitats and evolved strikingly different habits. In the genus *Dubautia* a

*One of the majestic silverswords,* **Argyroxiphium sandwicense** *subsp.* **macrocephalum** *('āhinahina) in Haleakalā Crater on Maui. Hybridization in the Hawaiian flora. The typical silversword growth pattern is seen — one stem, long, bayonet-type leaves, and one massive inflorescence.*

series of species can be seen that range in habit from *Dubautia scabra,* which forms mat-like clumps in lava fields on the island of Hawai'i, to larger, shrubby species, such as *D. linearis, D. ciliolata,* and *D. menziesii,* that are found in dry areas of Hawai'i and on Haleakalā on Maui. Several other species are found in dry to moist forests and are tree-like in habit. One species, *Dubautia waialealae* is found on Mt. Wai'ale'ale, the wettest area in Hawai'i. The dry land species all have adaptations for living in dry habitats, such as the presence of hairs, thick leaves with thick cuticles, or leaves that are reduced in area. What has also aided the exploitation of dry habitats is the evolution of differences in the structure of internal tissues. Those living in wet areas do not have such adaptations. One of the species is even a vine — *Dubautia latifolia.*

The genus *Argyroxiphium,* the silverswords, and the genus *Wilkesia,* the *iliau,* are closely allied to *Dubautia.* The Haleakalā silversword has perfected a rosette habit and leaf anatomy that allows the plant to prosper in the harsh environment presented in such an alpine desert as Haleakalā. The closely related green-swords are plants similar in habit to the silverswords but their leaves lack the silvery hairs that cover the surfaces of the leaves of the latter group and they are found in wet environments such as bogs.

*Wilkesia* is found only on Kaua'i. It is a rosette tree with leaves that top an unbranched woody stem usually between 2–3 m (6–10 ft) tall. This habit may have evolved in order to position the leaves of the plant above the shrubby vegetation in which it is found.

Work by Dr. Gerald Carr at the University of Hawaii has shown that the three endemic genera of this group can be hybridized and that spontaneous hybrids are found in nature. The obvious gross morphological differences between *Dubautia* and *Argyroxiphium,* so well demonstrated on Haleakalā on East Maui, belies the close genetic relationship between them revealed by the natural occurrence of hybrids. This is one of the several unique Hawaiian examples of radiation, speciation and adaptation and the one for which the greatest amount of information is presently available.

Among the most interesting groups of plants in Hawai'i is the native lobelioids. The fascination that the endemic species of this family has held for botanists and others in the Hawaiian Islands is considerable. Aside from the esthetics inherent in the flowers of the group and the often monocaulous or rosette habits of the plants, the lobelioids represent one of the best examples of plant-bird coevolution available on earth.

There are, or were, 6 endemic genera of lobelioids in the Hawaiian Islands: *Brighamia, Cyanea, Clermontia, Delissea, Rollandia* and *Trematolobelia.* The worldwide genus *Lobelia* is also found in Hawai'i. The lobelioids in general are shrubs or single-stemmed plants capped by rosettes of leaves that are often large and strap-like. The flowers are showy and white, greenish, purplish and shades in between and have floral tubes or corollas that are usually long and curved. The genus *Delissea* is nearly extinct with probably only one species out of 10 known still to be living. The species of the genus were adapted to dry forests and as that habitat and the birds that may have acted as pollinators were nearly gone by the time of contact it is likely that many more species existed at one time. The *Cyaneas* are still found

in Hawaiian wet, shady forests and most species also produce plants with single, unbranched stems. The range in size is considerable and juvenile forms of species of *Cyanea* often look quite different from the adult forms. Many of the species are now very rare and endangered. A number of species are very tall and palm-like in appearance. These are rare or thought to be extinct. The ultimate in this line of evolution to tall, palm-like habits is *Cyanea leptostegia* which is found on Kaua'i.

*Rollandia* is now reportedly restricted to O'ahu. It is very similar to *Cyanea* and several of the 9 species are now also very rare. The genus *Clermontia* is the most common of the lobelioids at present. All the species are shrubs or shrublike. They tend to cluster around natural openings in the wet forests where they are found and are therefore more liable to survive disturbance than their sister genera are.

The genera *Lobelia*, *Trematolobelia* and *Brighamia* are distinguished from the rest of the lobelioids in Hawai'i because they have retained what is probably the ancestral form of fruit that breaks open, or dehisces, when ripe. The others have fleshy fruit. However, in these genera

Opposite page: *A natural hybrid between the Haleakalā silversword and* **Dubautia menziesii**, *the* **na'ena'e**.
Below: *One of the species of* **na'ena'e**, **Dubautia menziesii**, *in Haleakalā Crater, Maui. This plant is a much-branched shrub with small leaves. The plant produces many inflorescences and does not die after flowering.*

the habit is unbranched stems with rosettes of leaves on top. *Brighamia* is certainly one of the weirdest looking plants known in Hawai'i. The one remaining species is found clinging to cliffs on the islands of Kaua'i and Moloka'i. The base of the plant is bulbous which allows it to rock in response to the impact of wind while it clings to narrow ledges.

This group of plants no doubt evolved with the family of birds called the Honeycreepers (Drepanididae). Members of this bird family known from the time of contact show evolutionary specialization based on feeding habits. One species, *Drepanidis funerea*, the *mamo*, showed the curved beak necessary for negotiating the long, curved floral tubes of the lobelioids. The species is now extinct. There may have been many more than this one species in this genus before man arrived in Hawai'i. The encouragement apparently offered by plants to birds and vice versa for mutual benefit was no doubt responsible for some of the fantastic examples of evolution hinted at by the remnants we have available today.

# EVOLVING LANDSCAPES

Nowhere else on this planet, other than, perhaps, in Iceland, can one see earth in the making as clearly as one can in Hawai'i. Recently, the eruptions at Kīlauea on the Big Island have been nearly continuous. Any visitor who is fortunate enough to see the eruptions that often consist of great fountains of lava, or even the less spectacular subdued flow of lava streams from rift zones, can see and

understand island building instantly. In recent historic times the only active volcanism has been on the Big Island but with Maui visible next door it is easy to visualize the process that has created the islands one after the other and will probably continue to do so long after man is gone. All of the native flora has probably had to learn to cope with this phenomenon and volcanism in the Hawaiian Islands has no doubt had significant effects on the evolution of the biota.

Aside from the creation of new land to be colonized — and this itself has probably favored plants that can do so — the lava flows on islands in active states of volcanism create islands of vegetation called *kipuka*. These *kipuka* create conditions in which populations are isolated and this begins the process called allopatic speciation — the geographical separation of populations that leads to the creation of new species. There are numerous *kipuka* on the island of Hawai'i of various sizes.

Several native Hawaiian plants have adapted themselves to colonizing recent lava flows. The ability of the *'ōhi'a lehua* (*Metrosideros*) to colonize recent lava, as well as nearly every conceivable Hawaiian habitat save the very dry, or very high is remarkable.

---

*A natural setting showing the contrasting growth forms of* **Argyroxiphium** *and* **Dubautia** *side by side in Haleakalā Crater on Maui.*

**Brighamia insignis**. *The "Cabbage-on-a-stick." This member of the Campanulaceae, the lobelioid group, is unique. The species is usually found clinging to cliff faces on the Nāpali Coast of Kaua'i. Here it is shown in cultivation in the Pacific Tropical Botanical Garden.*

# VEGETATION ZONES

Much of the Hawaiian landscape has been drastically changed by man. The average visitor sees mostly artificial landscapes. Even without man-made complications the Hawaiian Islands possess a highly diverse pattern of vegetation. This is due to diverse topography and rainfall patterns. Vegetation zonation in Hawai'i is based primarily on elevation and precipitation. Elevation determines temperature. The amount of rainfall in an area is determined by which side of a mountain one happens to be on or near. The trade winds blowing through most of the year from the northeast shed most of their moisture on the windward slopes or just over onto the leeward sides if the mountains are not too tall. The trades are moisture-rich because they have been blowing over thousands of miles of open sea before they reach Hawaiian shores. Depending on how high the mountains are and how many "wrinkles" there might be in the topography, local variation in rainfall could be very high. Visitors can literally see this phenomenon if they take the easy drive up the Pali Highway from

*Acacia koa* (**koa**). *One of the economically most important native Hawaiian trees.*

Honolulu on the island of O'ahu and stop at the well-known Pali Lookout. From a relatively new reinforced concrete platform they can look over a good part of the windward side of the island with the communities of Kāne'ohe and Kailua spread out below them. If it is a particularly windy day, the force of the winds blowing up and over the *pali*, or cliff, is very strong. The observant visitors will notice a great difference between the "slopes" on the windward side as opposed to those on the leeward side over which they have just driven. The windward slopes are sheer, for it is on this side that weathering and erosion has been greater. On a mature island like O'ahu it results in scenes of great beauty.

Age is also an important factor contributing to the variation of vegetation patterns in Hawai'i. Each of the islands has a unique topography corresponding to its age and the amount of erosion that has taken place. Patterns have developed based upon the particular plants that have been able to colonize a given island. Thus, species such as *Cryptocarya mannii* or *Elaeocarpus bifidus*, the only Hawaiian members of these two genera, occur only in the mixed mesophytic forests of Kaua'i and O'ahu. The more usual case, however, is that similar forests on different islands consist of representatives of the same genera, although usually not the same species.

It cannot be stressed enough that the problem that has contributed most to the difficulties of understanding natural vegetation patterns in the Hawaiian Islands is the severe degradation of native ecosystems by man. Because of these extensive alterations and the limited extent of the original native vegetation types it is now extremely difficult, and in some cases impossible, to construct a reasonable picture of the original ecosystems.

We will attempt to classify the Hawaiian vegetation into broad zones in order to convey information about the plants of Hawai'i.

## I. STRAND

Coastal sites affected by salt spray, seawater and limited by dunes and coral outcrops. Strand vegetation is dominated primarily by low shrubs and perennial herbs. The species commonly seen are *Scaevola*

Below: *Misty ridges of the Ko'olaus on O'ahu above Punalu'u.*

Opposite page: *The map shown here is modeled after that by Ripperton & Hosaka (1942) as modified by the Department of Agronomy and Soil Science at the University of Hawaii. We have shown our vegetation zones in a general way. The strand zone is represented essentially by the outlines of the islands.*

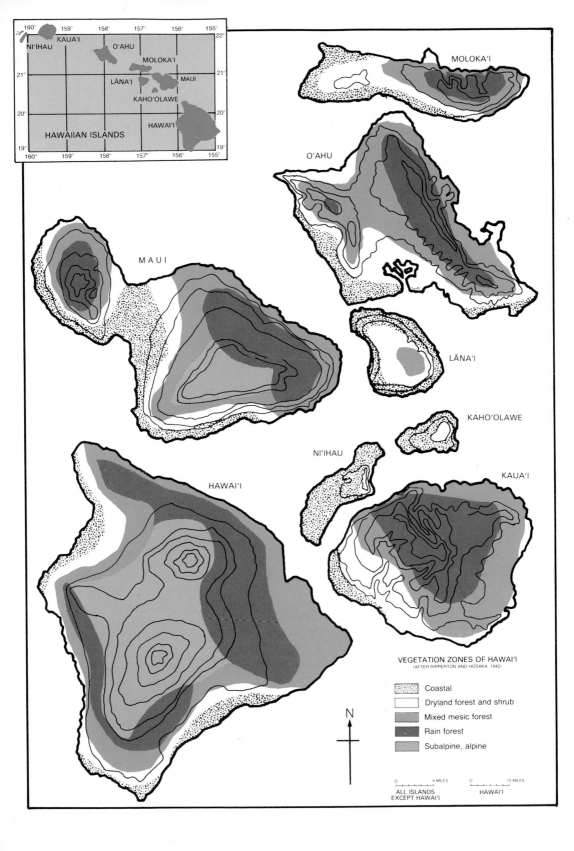

160° 159° 158° 157° 156° 155°

NI'IHAU
KAUA'I
O'AHU
MOLOKA'I
LĀNA'I    MAUI
KAHO'OLAWE
HAWAI'I
HAWAIIAN ISLANDS

MOLOKA'I

O'AHU

MAUI

LĀNA'I

KAHO'OLAWE

NI'IHAU

HAWAI'I

KAUA'I

**VEGETATION ZONES OF HAWAI'I**
(AFTER RIPPERTON AND HOSAKA, 1942)

Coastal

Dryland forest and shrub

Mixed mesic forest

Rain forest

Subalpine, alpine

N

0    6 MILES        0        10 MILES

ALL ISLANDS          HAWAI'I
EXCEPT HAWAI'I

*sericea\*, Jacquemontia ovalifolia\*, Sesuvium portulacastrum\*, Sida cordifolia\*, Vitex rotundifolia\*, Heliotropium anomalum\*, H. curassavicum\**, and the grass *Sporobolus virginicus\**. All of these plants are relatively easily dispersed and are relatively widespread at least in the Pacific Basin. Some have floatable fruits such as *Scaevola sericea* which has undoubtedly colonized Hawaiian shores many times. Other species such as those of *Boerhavia* have sticky fruits which adhere easily to birds. Approximately 25% of the indigenous, non-endemic species in Hawai'i occur in strand vegetation including, in addition to those mentioned above, *Portulaca lutea\*, Lycium sandwicense\*, Ipomoea pescaprae\*, I. stolonifera\*, Lepturus repens\** (Leeward Islands only), *Tribulus cistoides\*, Bacopa monnieria\**. Some indigenous strand species are very rare and may represent recent colonizations that have not spread extensively, such as *Cressa truxillensis\**. Others were once more common, such as *Lysimachia mauritiana\**, historically known from Ni'ihau, Moloka'i, Maui and Hawai'i, but this century collected only a few times primarily from remnant populations on Moloka'i.

There are also endemic species in strand communities, such as *Nama sandwicensis* and *Chenopodium oahuense*. Many of these endemic strand species were derived from upland ancestors that occurred in dryland forest or rain forest including *Chamaesyce degeneri, Lipochaeta integrifolia, Scaevola coriacea, Phyllostegia variabilis*, and *Achy-*

*ranthes atollensis*. The only trees in this zone are naturalized species, some of which, such as *Calophyllum inophyllum*, were introduced by the Polynesians.

Strand communities in Hawai'i are now mere remnants of their former selves, primarily because of intensive development of the shores for tourism. Many areas have been invaded by mangrove (*Rhizophora mangle*) along the shores of many of the main islands. Other trees, such as *Terminalia catappa* and *Tournefortia argentea*, are now common constituents of the strand. Many halophytic grasses, such as *Cynodon dactylon*, and herbs, such as *Atriplex semibaccata*, are now

*An example of a coastal habitat. Mo'omomi Beach, northwestern Moloka'i.*

*indicates an indigenous species that is not endemic. Most indigenous plants in Hawai'i are endemic, and these appear without asterisks as do also introduced plants.

42

more common than the native species. Some of the rarest plants in Hawai'i, such as *Lysimachia mauritiana* or *Scaevola coriacea*, and extinct species like *Phyllostegia variabilis* or *Achyranthes atollensis* (both from the northwestern Hawaiian Islands), were members of the strand community. *Halophila hawaiiana* is the only flowering plant found in tidal flats and should perhaps be counted as a member of this zone.

## II. COASTAL

The coastal zone as herein designated is not influenced by salt spray or seawater. There are arid and mesic areas of this zone.

The arid parts of this zone are found mostly on the leeward sides of the islands. The rainfall of this zone varies from 15–40 inches per year and the elevation ranges from sea level to 300 m. Coastal and strand vegetation are the *only* habitats found on the northwestern Hawaiian Islands (the leeward islands from beyond Kaua'i to Kure Atoll). A major portion of the rain of the zone falls during the winter months. This pattern of winter rain is called Kona weather in Hawai'i as the prevailing winds during this short period of time are from the south. This is one of the most completely altered of the vegetation zones and it is therefore difficult to say what it was really like before man.

Some of the principal tree components of the arid coastal zone are *Erythrina sandwicensis*, *Santalum ellipticum*, and *Myopor-*

um sandwicense*. The latter species has an extremely wide range in most arid or mesic habitats and at high and low elevations.

Some common species of this arid coastal zone include Waltheria indica*, Sida fallax*, Capparis sandwichiana, Caesalpinia bonduc*, Sesbania tomentosa (rare), Cassytha filiformis*, Plumbago zeylanica*, and Osteomeles anthyllidifolia*, Gossypium sandvicense, Panicum torridum, P. fauriei, and the now relictual fern Marsilea villosa (O'ahu and Moloka'i).

Most coastal habitats are now covered with alien vegetation composed of Leucaena leucocephala and Prosopis pallida scrub. Leucaena, commonly called koa-haole, forms dense impenetrable thickets in many areas. Other common naturalized species include Merremia aegyptia, Chenopodium murale, grasses such as Rhynchelytrum repens, Chloris inflata, and Cynodon dactylon, composites such as Bidens pilosa and B. alba, and Tridax procumbens. Cordia subcordata, the Polynesian introduction known as kou, is also found in this zone as well as the nearly ubiquitous coastal plant, Casuarina equisetifolia, introduced after contact.

The mesic coastal areas are obviously wetter than the arid coastal areas, and are found on the windward sides of the islands. In these areas there existed a mesic forest which has been virtually eliminated. It was dominated by hala, Pandanus tectorius*. The hau, Hibiscus tiliaceus*, along with two naturalized Polynesian introductions, Thespesia populnea, and Aleurites moluccana, were the dominants in these areas during the past 1,000–2,000 years. These areas now include common species

such as Passiflora laurifolia, Furcraea foetida, which reproduces by bulbils, and three other Polynesian introductions, Tacca leontopetaloides, Cordyline terminalis, and Morinda citrifolia, and the common guava, Psidium guajava. Some native species still occur in these areas such as the fern, Ophioglossum concinnum.

## III. DRYLAND FOREST AND SHRUB

This was at one time a diverse vegetation, but there is little of this remaining in Hawai'i. Most of the area that would have once been covered with this vegetation is now largely given over to grazing of cattle and sugar and pineapple agriculture. These areas may have been extensively burned during the Polynesian period. Dryland forest ranged from about 200–300 m to about 900 m in elevation. It is difficult to define upper and lower elevational limits of this vegetation. It occurs primarily on the leeward sides of the main islands, but includes much of the smaller islands of Lāna'i and probably all of Kaho'olawe. In the past these last two islands as well as dry forest areas of East Maui were wetter. The removal of much forest on Haleakalā has resulted in a drier climate over this whole area. A present problem is that many native species that produce fruit are prevented from establishing seedlings because of the aggressive growth of grasses such as African fountain grass, Pennisetum setaceum. Introduced rats also rapidly devour many seeds as they are produced. Seed-boring insects also make an end to many seeds. There are relict areas on O'ahu such as in the Wai'anae Mountains, but these remnant areas may eventually be entirely eliminated by fire. A particularly rich remnant in the Wai'anae Mountains consists of a small area of only a few hundred square meters that is completely surrounded by burned slopes with no native species.

About 22% of the total indigenous species in Hawai'i occur in this zone, but there are only several endemic species. Many of the dry forest species are only seen in relictual areas like Auwahi, Maui. Other examples are the endemic species of *Gouania*, now only known from a few hundred individuals.

Some of the common species are *Erythrina sandwicensis*, *Diospyros sandwicensis* (often dominant), *Reynoldsia sandwicen-* sis, *Nothocestrum* spp., *Rauvolfia sandwicensis*, *Canthium odoratum**, *Ochrosia compta*, and *O. haleakalae*. *Santalum paniculatum* is common only on the island of Hawai'i in this zone. There are also *Planchonella sandwicensis*, *Nesoluma polynesicum**, *Caesalpinia kauaiense* (endangered), and *Colubrina oppositifolia*. *Sapindus oahuensis* is the only simple leaved species in its genus. Also common are *Eugenia reinwardtiana**, *Argemone glauca*, *Acacia koa*, *Cocculus trilobus**, and *Myoporum*. *Myrsine lanaiensis* is also found in this zone although the other members of the genus in

**Reynoldsia sandwicensis** *(Araliaceae).* **'Ohe** *or* **'ohe-kukuluāe'o***. A genus of 6 or 7 species of dry forest trees; native to Samoa, the Marquesas and the Society Islands as well as Hawai'i.*

Hawai'i are more likely to be found in wet forest or bogs.

Two species of *Nototrichium* occur in this zone. *Nototrichium humile* is now known only from a few relictual populations in the Wai'anae Mountains and from one population in Auwahi, East Maui. The large endemic Hawaiian genus *Lipochaeta* is well adapted to this zone. Ten species of this genus occur here. One, *L. venosa*, known only from a series of cinder cones on leeward Hawai'i, is listed as endangered. *Tetramolopium* is another genus with unique Hawaiian species. There are 11 species known primarily from low or high elevation dry habitats. The four dry forest species are now all thought to be extinct.

The dryland forests and shrublands are now populated mainly by aggressive species such as *Lantana camara*, molasses grass (*Melinis minutiflora*), *Pennisetum setaceum* and *P. ciliare*. Other common species include *Prosopis pallida*, *Leucaena leucocephala*, and *Opuntia ficus-indica*.

## IV. MIXED MESIC FOREST

This is the most species-rich of the vegetation zones in the islands, occurring from about 750 to 1,250 m. It is best developed on the Kōke'e Plateau, Kaua'i. This vegetation is now highly disturbed or entirely eliminated in many areas, but it probably originally occurred on all of the main islands except the smallest. Even Kaho'olawe, however, may have had some elements of mesic forest. The richest assemblage of species occurs on the older islands of O'ahu and especially Kaua'i. These areas have less

rainfall than the rain forests, but do not suffer extended dry periods. Formerly they were open-canopy forests consisting of a diverse mixture of trees and shrubs. It is often difficult to clearly demarcate where the dry forest ends and mixed mesic forest begins. Along with rain forests, mixed mesic forests harbor the majority of endemic species in virtually all of the larger genera, such as *Cyanea*, *Schiedea*, and *Pelea*. The most notable absences are members of *Cyrtandra*, *Phyllostegia*, and *Stenogyne*, which occur predominantly in the wetter rain forest habitats. Two of the most common trees of this vegetation type are the ubiquitous *Acacia koa* and *Metrosideros polymorpha*. The striking feature of the mixed mesic forests is that mixed with these two trees, that otherwise dominate in so many habitats in Hawai'i, are a relatively large number of other tree or shrub species. Some of these are *Antidesma platyphyllum*, *A. pulvinatum*, *Claoxylon sandwicense*, *Alphitonia ponderosa*, *Strebulus pendulinus\**, *Charpentiera* (5 spp.), *Chamaesyce* (several spp.), *Ochrosia* (4 spp.), *Nestegis sandwicensis*, *Pittosporum* (several spp.), *Santalum freycinetianum*, *Eugenia reinwardtiana*, *Wilkesia* (2 spp., Kaua'i only), *Psychotria* (several spp.), and *Elaeocarpus bifidus* (O'ahu and Kaua'i). Other elements are less common or even very rare, such as members of the endemic genera *Kokia* (4 spp.) and *Hibiscadelphus* (6 spp.). The species of these two genera occur in mixed mesic and dry forests, and now are only relicts. Several of the *Hibiscadelphus* species

---

*Mixed mesic forest.* **Cyanea leptostegia**, *a lobelioid member of this zone.*

apparently are extinct and others, such as *H. hualalaiensis* and *H. distans*, are near extinction. Similarly, *Kokia cookei* is now known only from grafted stems in cultivation, *K. drynarioides* of the dry forest of Hawai'i Island is endangered, *K. kauaiensis* is relatively uncommon in a few gulches on the Kōke'e Plateau, Kaua'i, and *K. lanceolata*, formerly of O'ahu, is extinct. Mixed mesic forests harbor the largest numbers of unique and rare Hawaiian plants, such as *Remya* (3 spp.), *Isodendrion* (3 spp. mixed mesic, 2 spp. dry forest), *Delissea* (10 spp., 7-9 now extinct), and the monotypic *Munroidendron* (Kaua'i) and *Neowawraea*.

Other species include *Bonamia menziesii*, which is a rare member of the morning glory family, *Tetramolopium lepidotum*, and *T. filiforme*, that are both found in the Wai'anae Mountains, O'ahu, and are among the few species out of the 11 described for the genus that still survive, *Lepidium serra* is found in mixed mesic forest on Kaua'i, and *L. arbuscula* is a woody mustard from O'ahu. There are 6 species of *Neraudia* found in this zone as well as *Phyllanthus distichus*, and *Alyxia oliviformis* (which is a common understory vine). *Bidens cosmoides* is a composite with giant heads with exserted flowers from Kaua'i. It is probably bird-pollinated. *Rumex giganteus*, which is a large liana or shrub from Maui, Moloka'i, and Hawai'i, as well as *R. albescens* from O'ahu, Kaua'i, and Nihoa, are also found in this zone. *Nothocestrum peltatum* is a recently rediscovered member of this endemic genus from mixed mesic forest and three species of *Dubautia* are also found here: *D.*

*herbstobatae*, *D. latifolia* (which is the only liana in the genus and is from Kaua'i) and *D. laevigata*. Four species of *Lipochaeta* are found here: *L. subcordata*, *L. micrantha*, *L. tenuis*, and *L. tenuifolia*. Several other species that bear mentioning are *Strongylodon ruber*, *Caesalpinia kauaiense* (which is endangered), *Luzula hawaiiensis*, *Dianella sandwicensis\**, *Wikstroemia furcata*, and *W. oahuensis*.

The mixed mesic forests, although existing in some areas in much the same form as they probably did before man, have been seriously degraded in the past 100 years. Large areas such as on the slopes of Haleakalā have been converted to pastures of kikuyu grass (*Pennisetum clandestinum*). This aggressive African grass, which forms such im-

penetrable cover as to prevent germination of the native plants, has invaded many areas adjacent to pastureland. Another very serious weed is *Passiflora mollissima*, banana poka, a rapidly growing vine that smothers the plants that it covers. It is a difficult problem on, as yet, only Kaua'i and Hawai'i. Many other alien species grow in these mixed mesic forests, forming extensive and dense growth, including *Schinus terebinthifolius*, *Persea americana* (avocado), *Corynocarpus laevigatus*, *Myrica faya*, *Rubus argutus*, *Olea europaea*, *Lantana camara*, *Grevillea banksii*, *G. robusta*, *Macaranga* spp., *Tibouchina urvilleana*, and *Melastoma candida*.

A vegetation type that is difficult to place occurs in low elevation mesic valleys. This area is dominated by species such as *Pisonia* and *Aleurites moluccana*. In some ways it is an extension of the vegetation type found on windward coastal sites, and like those areas these also have been largely replaced with alien plants. It once harbored species of *Cyrtandra* such as *C. grayi*, *C. filipes*, *C. hawaiiensis*, *C. confertiflora*, *C. spathulata*, and *C. laxiflora*. It now includes *Syzygium jambos*, *S. malaccense*, *Mangifera indica*, *Trema orientalis*, *Schefflera actinophylla*, *Psidium cattleianum*, and *P. guajava*.

## V. RAIN FOREST

The Hawaiian rain forests

*Mixed mesic forest.* **Kokia cookei (koki'o)** *may well be the rarest plant in the Hawaiian Islands.*

are characterized by high rainfall that usually ranges from 150 to 300 or more inches per year. During the winter months the rain forests are often enveloped in clouds for long periods of time. Periods of heavy rain are interrupted by periods of mist. During the remainder of the year the morning hours are often clear. The afternoons and nights, however, are usually given to clouds and rain. Hawaiian rain forests occur at elevations from about 450 to 1,700 m (ca. 1,350–5,100 ft). They are found above mixed mesic forests. Rain forest is probably the first native vegetation visitors now encounter in many areas of Hawai'i. In areas below 600 m in the Ko'olau Mountains, O'ahu, for example, little original vegetation exists, but on the ridges, highest gulches, and along the summit some rain forest remains. The Hawaiian rain forest is dominated by 'ōhi'a (Metrosideros polymorpha), but koa (Acacia koa) is also very common.

The forest presents a dense tangle of branches and rotting logs, which often makes movement difficult. Along with mixed mesic forests, rain forests contain the greatest diversity and largest proportion of unique species. Common shrubs and trees include the tree fern, Cibotium spp., Tetraplasandra (6 spp.), especially T. oahuensis, Cheirodendron trigynum, several other species of Cheirodendron found only on Kaua'i, Pelea clusiifolia, Broussaisia arguta, Syzygium sandwicensis, Ilex anomala*, Myrsine sandwicensis, M. lessertiana, Rubus hawaiiensis, and Perrottetia sandwicensis. Vines include Smilax melastomifolia and Alyxia oliviformis. Epiphytic vegetation is com-

mon, consisting primarily of mosses such as Acroporium fuscoflavum, Macromitrium owahiense, Leucobryum gracile, Rhizogonium spiniforme, and Racopilum cuspidigerum, and numerous ferns such as Adenophorus montanum, A. hymenophylloides, Mecodium recurvum, Spaherocionium lanceolatum, Ophioglossum pendulum, Asplenium nidus, Elaphoglossum hirtum, and Grammitis hookeri. A number of epiphytic flowering plants are also present, including Peperomia tetraphylla*, Astelia menziesiana, Freycinetia arborea, Cheirodendron trigynum, and the rare Clermontia peleana.

A great many of the species of the larger native genera occur in the rain forest, including Phyllostegia (27 total native species), Stenogyne (20 spp.), Labordia (17 spp.), Cyrtandra (15 spp.), Clermontia (23 spp.), and Cyanea (50 spp.). Myrsine with 21 native species has a great number of them in rain forests, but nearly all of them are endemic to Kaua'i. The rain forest also harbors the three species of orchids endemic to Hawai'i, Anoectochilus sandvicensis, Liparis hawaiensis, and Platanthera holochila, the latter of which has its nearest relatives in Alaska.

Other rain forest species include: Gunnera petaloidea found on steep slopes, Adenostemma lavenia*, Pipturus albidus, and Touchardia latifolia, which is a monotypic endemic genus. Urera sandvicensis and Pilea peploides* are rain forest inhabitants as are Lobelia gaudichaudii, and Rollandia st.-johnii. The latter two are found in what has been called the cloud forest region. Trematolobelia is an

Rain forest. Dense numbers of epiphytes cover trees in the Waikamoi area of East Maui.

endemic genus with 4 species found on windswept ridges. Three species of *Scaevola* are found in rain forest: *S. glabra, S. chamissoniana,* and *S. gaudichaudiana. Phytolacca sandwicensis,* and three species of *Chamaesyce* are also found here: *C. clusiifolia, C. remyi,* and *C. rockii.* These three are examples of adaptive radiation into wet forests from small-leaved ancestors in drier habitats.

Hawaiian rain forests are perhaps not as altered as other native ecosystems, but they are the habitats that are being most adversely impacted at the present time. The problem is primarily due to invasion by feral animals, especially pigs that root up plants like *Astelia menziesiana,* lobelioids, and tree ferns. In areas opened up by pigs, weeds, such as guava and blackberry, quickly replace native species. One of the most serious alien plants in rain forests is *Clidemia hirta,* which over a period of 40 years has become the dominant understory plant in the Ko'olau Mountains, O'ahu. It is now repeating this kind of invasion on other islands, including Kaua'i, Maui, and Hawai'i.

## VI. BOGS

Hawaiian bogs form in relatively level montane areas where rainfall exceeds drainage. They are usually underlain by a layer of light gray impervious clay. The largest bog in Hawai'i is the Alaka'i Swamp on the summit plateau of Kaua'i which is underlain by denser, less porous lavas that resist erosion. Other bog areas include Mount Ka'ala, O'ahu, Pēpē'ōpae Bog, Moloka'i, Pu'ukukui and Mt. 'Eke bogs of

West Maui, smaller bogs on East Maui, and the summit area of the Kohala Mountains, Hawai'i.

The vegetation of Hawaiian bogs generally consists of irregular hummocks of cushion-like, low shrubs, sedges, and grasses. Occasionally, larger shrubs or even small trees occur, especially along the bog margins or on raised hummocks. The dominant bog species form tussocks such as the sedges *Oreobolus furcatus* and *Rhynchospora lavarum.* Dwarf grass species such as *Dichanthelium isachnoides, D. cynodon, D. koolauense,* and *Deschampsia australis* are also common. The woody component of bogs consists primarily of species that are common in other vegetation zones but exist in dwarfed form in bogs. The most prominent species are 'ōhi'a (*Metrosideros polymorpha*), and *Vaccinium* spp. Other common constituents include the sedge *Machaerina angustifolia,* ferns such as *Sadleria* spp., the mosses *Rhacomitrium lanuginosum* and *Sphagnum palustre* (only in the Kohala Mountains, Hawai'i and Mt. Ka'ala, O'ahu) and the fern ally, *Lycopodium cernuum.*

Other species include *Acaena exigua,* which is very rare and found only on West Maui and Kaua'i, *Argyroxphium caliginis,* a member of the silversword alliance found only on West Maui, *A. grayanum,* found along bog margins on West Maui and in wet forest on East Maui, as well as *A. kauense* found on the island of Hawai'i where it is badly affected by alien moufflon sheep. *Dubautia imbricata,* in the

*Rain forest. Lush fern forest with* **hāpu'u** (*Cibotium sp.*) *in Hawai'i Volcanoes National Park.*

tarweed alliance along with the silverswords, is found only on Kaua'i. There are four other species of this genus found also in bogs: *D. paleata, D. raillardioides, D. waialealae* and *D. laxa.* The first three are found only on Kaua'i. The only two herbaceous Hawaiian violets are found in bogs (all other Hawaiian violets are shrubs): *Viola mauiensis* and *V. kauaensis.* Three species of the genus *Lobelia* are found in Hawaiian bogs: *L. kauaensis, L. villosa,* and *L. gaudichaudii,* as well as a geranium, *Geranium humile.*

Until recent decades, Hawaiian bogs remained relatively pristine. However, they are becoming increasingly degraded by the activities of man and his introduced animals. Trampling by hikers creates areas of mud and standing water not found in the natural state. The most serious problem is the relatively recent invasion of bog ecosystems by feral pigs. Pigs uproot bog plants to eat the succulent roots or copious earthworms. They seem to particularly enjoy eating the hearts of tree ferns and the flowers of endemic lobelioids. Fencing projects help to some degree, but pigs must be removed from these fragile Hawaiian ecosystems if there is to be any hope at all for their continued existence.

# VII. SUBALPINE WOODLAND, SHRUBLAND, AND DESERT

At elevations above 6,000 ft. on Hawai'i and Maui there are areas with a climate that is more temperate than tropical. Temperatures are lower, rainfall is less, and snow falls at the highest elevations during the winter months. Much of this area is above the so-called inversion layer — the area where humidity and temperature change dramatically. Parts of the subalpine zone are covered with open woodland, *mamane-naio* (*Sophora chrysophylla* and *Myoporum sandwicense*\*). Other areas have a relatively open shrubland such as the *Chenopodium oahuense* shrublands at Pohakuloa, in the saddle between Mauna Loa and Mauna Kea, Hawai'i. There are also more mixed shrublands, such as on the slopes of Haleakalā, that contain *Dubautia menziesii, Vaccinium reticulatum, Coprosma montana, C. ernodeoides,* and *Styphelia tameiameiae*\*. Other areas support mixtures of species such as *Santalum haleakalae,* a shrubby geranium (*Geranium cuneatum*), and *Dodonaea viscosa*\*. There are also patches of alpine grassland dominated by tussock-forming species such as *Deschampsia australis, Panicum tenuifolium,* and *Trisetum glomeratum.* These grasslands have been very seriously damaged by overgrazing and only very small areas are now in relatively pristine condition. On some of the highest parts of Maui and Hawai'i there are drier habitats that are often referred to as alpine deserts, harboring plants such as the famous silversword, *Agryroxiphium sandwicense,* as well as *Dubautia arborea, Silene struthioloides,* and *Tetramolopium humile.*

Other species of this region include *Haplostachys haplostachya* that was formerly found on Kaua'i, Moloka'i, and Hawai'i, but is now restricted to one pop-

*Sadleria squarrosa* ('apu'u), *an endemic fern found on wet cliff faces.*

ulation in the Pōhakuloa military training area of Hawai'i. *Sanicula sandwicensis, Artemisia mauiensis, Bidens micrantha,* and four species of *Dubautia* — *D. ciliolata, D. scabra, D. platyphylla,* and *D. reticulata* — are found vying for space in the drier areas of this zone. *Gnaphalium sandwicensium, Tetramolopium consanguineum, Lobelia grayana* are three more endemic species found here as was the now extinct *Silene degeneri. Schiedea haleakalensis* is found on cliff faces of this zone. An endemic spurge, *Chamaesyce olowaluana,* and the bird-pollinated *Geranium arboreum,* which has zygomorphic flowers, a feature alien to most geraniums, are also to be found here. Two endemic mints, *Stenogyne micrantha* and *S. rugosa* are found here also.

## VIII. CLIFFS

Cliff habitats are largely azonal, however, the majority of cliff sites occur on the windward sides of most of the main islands. They usually consist of fluted, wet basalt up to 1,000 m high, with essentially no soil and sparse vegetation. What vegetation exists consists of dwarf shrubs such as 'ōhi'a (*Metrosideros polymorpha*) and 'ōhelo (*Vaccinium* spp.). However, there are a number of species endemic to this habitat, including the bizarre *Brighamia,* a genus with a short conical fleshy stem and terminal cluster of leaves reminiscent of a head of cabbage. *Brighamia* apparently is an ancient member of the Hawaiian flora and botanists so far have no idea of what its ancestor was.

Other cliff endemics include *Peucedanum sandwicense,* found on Kaua'i (K), Moloka'i (Mo) and Maui (M), *Artemisia australis,* found on all the main islands, 3 species of ko'oko'olau, *Bidens hillebrandiana* (Mo, M, H),* *B. mauiensis,* (L, M, Ka) and *B. molokaiensis* (O, Mo). A species of *Lobelia* (*L. niihauensis*) is found only on cliffs of Kaua'i. Other endemics include *Hedyotis st.-johnii* (K), and 4 species of the genus *Schiedea: S. apokremnos* (K), *S. globosa* (O, Mo, M), *S. stellarioides* (K), and *S. verticillata* (Nihoa).

## IX. EXOTIC LANDSCAPES

Unfortunately, the average visitor to Hawai'i rarely sees a native plant. If the visitor restricts himself to the coastal area resorts it is likely that he will not see a

---

*See page 59 for island acronyms.
*Haleakalā, East Maui, is a good example of an alpine desert.*

single native plant. Much of the flora of coastal or lowland areas is exotic. They have been brought in accidentally or purposefully by man. In Kapi'olani Park near Waikīkī, for example, kiawe (*Prosopis pallida*) and ironwood (*Casuarina equisetifolia*) trees make up nearly all of the tree life found there. Even the grass underfoot is not native. Much of the color among roads and around houses is due to ornamental species of *Bougainvillea*, *Ixora*, *Poinsettia*, and other exotics. Only when the visitor goes to higher elevations does the proportion of native species increase. Hawai'i Volcanoes National Park on the Big Island, Haleakalā on East Maui and Kōke'e State Park on Kaua'i are three areas which are easy to reach by automobile and in which native plants will be seen. In Kōke'e there are also self-guided trails available. The visitor should take the opportunity while in Hawai'i to try to see some examples of native forest, for that experience is one that fewer and fewer of us will have as time goes by. At very least, a visit to several of our botanical gardens and arboreta would be worthwhile. The Waimea Arboretum on O'ahu has a special garden of rare and endangered Hawaiian plants, and the City and County of Honolulu has a native Hawaiian plant area in Ho'omaluhia Park in Kane'ohe. There are other gardens and arboreta that can be visited as well.

*Overleaf:* **Trematolobelia macrostachys** *(Campanulaceae)* **Koli'i**. *The flowers are borne on a circle of branches terminating the stem.*

## CODE FOR CAPTIONS

I     :   indigenous
Eg    :   endemic genus
Es    :   endemic species
A     :   annual
HP    :   herbaceous perennial
V     :   vine
S     :   shrub (definitely woody)
T     :   tree
RS    :   rosette shrub (unbranched or
          sparingly branched), shrub
          with a terminal rosette of leaves.
RT    :   rosette tree (unbranched or
          sparingly branched), tree with a
          terminal rosette of leaves.
P     :   parasite

# Plants and Flowers

HI : Hawaiian Islands (Kaua'i,
Ni'ihau, O'ahu, Moloka'i,
Lāna'i, Kaho'olawe, Maui,
Hawai'i) inclusive.
LI : Leeward Islands
K : Kaua'i
N : Ni'ihau
O : O'ahu
Mo : Moloka'i
M : Maui (Me – East Maui, Mw –
West Maui)
Ka : Kaho'olawe
L : Lāna'i
H : Hawai'i

The number after the family and status
designation indicates the number of
native Hawaiian species in the genus.

59

# STRAND

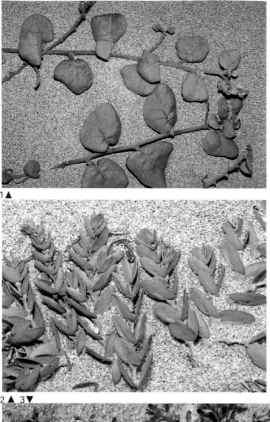

1. **Solanum nelsoni** (*Solanaceae, Es, 3 or 4, HP, LI and HI except L, Ka*). *A strand plant known, but rare, from most of the major islands and common on some of the Leeward Islands.*

2. **Chamaesyce degeneri** (*Euphorbiaceae, Es, 14, S, HI except L & Ka*). **'Akoko**. *Prostrate to decumbent subshrub scattered in coastal and strand vegetation on all of the main islands except Lāna'i.*

3. **Nama sandwicensis** (*Hydrophyllaceae, Es, 1, A, LI, HI except Ka*). *An uncommon plant known from most of the major islands and from Lisianski and Laysan Islands of the Leeward Islands. It is perhaps a short-lived perennial rather than an annual plant.*

4. **Solanum nelsoni.** *Close-up of the flower.*

5. **Heliotropium anomalum** (*Boraginaceae, I, 2, HP, LI, HI except L, and Ka*). **Hinahina**. *A common, decumbent perennial with silvery leaves, forming mats on sand.*

4▲ 5▼

6. **Sesbania tomentosa** *(Fabaceae, Es, 1, HP, S, HI except N, also on Nihoa & Necker).* **'Ohai.** *This plant is from southwestern O'ahu.*

7. **Sida cordifolia** *(Malvaceae, I, 4, S, HI).* **'Ilima.** *One of many species occuring in the islands whose taxonomy is poorly understood.*

8. **Gossypium sandvicense** *(Malvaceae, Es, 1, S, HI except H).* **Ma'o.** *Scattered along the coasts of all of the major islands except the Big Island.*

9. **Plumbago zeylanica** *(Plumbaginaceae, I, 1, S, HI).* **'Ilie'e.** *A common straggling shrub in coastal and other low, dry areas.*

6▲ 7▼

8▼ 9►

10. **Pandanus tectorius** *(Pandanaceae, I, 1, T, HI).* **Hala**. *This species is common in coastal areas and lowlands. It was an important economic plant for the Hawaiians.*

11. **Schiedea globosa** *(Caryophyllaceae, Eg, 22, HP, O, Mo, M).* **Mā'oli'oli**. *An uncommon herb from coastal habitats.*

12. **Scaevola sericea** *(Goodeniaceae, I, 10, S, LI, HI).* **Naupaka**. *A common white-fruited shrub strand plant ranging throughout the coasts of the tropical and subtropical Pacific and Indian Oceans.*

13. **Scaevola coriacea** *(Goodeniaceae, Es, 10, S, HI except Ka, H).* **Naupaka**. *An almost extinct species presently known from Maui and a few small islets off Maui and Moloka'i, but formerly on most of the main islands.*

10▲ 11▼

12▲  13▼

14. **Capparis sandwichiana** (*Capparaceae, Es, 1, S, HI and Midway Atoll, Pearl & Hermes Atoll, Laysan Island*). **Maiapilo**. *Low-growing shrub of Hawaiian coasts with nocturnal flowers and cucumber-like fruits.*

15. **Brighamia insignis** (*Campanulaceae, Eg, 2, RS, N, K*). **Alula**. *A rare stem succulent growing on steep vertical cliffs above the ocean. Another species is found on Moloka'i, Lāna'i, and Maui.*

16. **Vitex rotundifolia** (*Verbenaceae, I, 1, S, HI except Ka*). **Pōhinahina** *or* **kolokolo-kahakai**. *Common coastal shrub with fragrant flowers.*

14▲ 15▼  16▶

17. **Jacquemontia ovalifolia** *subsp.* **sandwicensis** *(Convolvulaceae, Es, 1, HI).* **Pā'ū-o-hi'iaka.** *The subspecies is endemic, but the species is indigenous. It is a common strand plant on all of the islands.*

18. **Tetramolopium rockii** *(Asteraceae, Es, 11, S, Mo). A prostrate, small, compact shrub known only from the lithified sand dunes in coastal scrub of northwestern Moloka'i.*

19. **Chamaesyce celastroides** *var* **kaenana** *(Euphorbiaceae, Es, 14, S, O).* **'Akōko.** *A rare and endangered beach plant known only from Ka'ena Point, O'ahu.*

17▲ 18▼                                                    19▶

20. **Colubrina oppositifolia** (*Rhamnaceae, Es, 2, T, O, H*). **Kauila**. *A rare tree with extremely hard wood highly valued by the Hawaiians.*

21. **Rauvolfia sandwicensis** (*Apocynaceae, Es, 1, S or T, HI except Ka*). **Hao**. *A small tree whose fruits are purplish-black at maturity.*

22. **Dodonaea eriocarpa** (*Sapindaceae, Es, 2, S, HI except N & Ka*). **'A'ali'i**. *Common shrub in dry and mesic forests.*

23. **Erythrina sandwicensis** (*Fabaceae, Es, 1, T, HI*). **Wiliwili**. *A common gnarled tree in lowland dry areas throughout the islands. Flower color varies from orange to ivory or even greenish.*

20▲

21▲  22▼

23►

24. **Scaevola kilaueae** (*Goodeniaceae*, Es, 10, S, H). **Huahekili-uka**. *A local plant from the dry ash-fields southeast of Kīlauea.*

25. **Hibiscus brackenridgei** *var.* **mokuleiana** (*Malvaceae*, Es, 6, S, O). **Ma'ohauhele**. *A rare shrub from the Wai'anae Mts. of O'ahu; other forms are on Lāna'i and Maui.*

26. **Planchonella auahiensis** (*Sapotaceae*, Es, 6, T, M). **'Āla'a**. *A small-flowered tree easily recognized by its yellow fruits.*

27. **Haplostachys haplostachya** (*Lamiaceae*, Eg, 5, HP, H). **Honohono**. *An almost extinct genus of Hawaiian mints.*

24▲ 25▼

26▼ 27►

28. **Hedyotis formosa** (*Rubiaceae, Es, 20, S, Mw*). *A low shrub on rocky walls of deep gulches.*

29. **Myoporum sandwicense** (*Myoporaceae, I, 1, S or T, HI except Ka*). **Naio**. *A common shrub or tree of dry areas with white or lavender flowers.*

30. **Styphelia tameiameiae** (*Epacridaceae, I, 1, S, HI except N & Ka, where it may have occurred in the past*). **Pūkiawe**. *A common shrub of dry or mesic areas, rain forests, alpine scrub, and bogs. It occurs at elevations from 60 to 3,230 m. The species has either red or white fruit.*

28▲ 29▼                                                                          30►

31. **Caesalpinia kauaiense** *(Fabaceae, Es, 1, T, K, O, M, H).* **Uhiuhi**. *A rare, almost extinct tree with purplish flowers.*

32. **Nototrichium sandwicense** *(Amaranthaceae, Eg, 2, S or T, HI).* **Kulu'ī**. *A common shrub or small tree on all of the major islands.*

33. **Nothocestrum breviflorum** *(Solanaceae, Eg, 4, T, H).* **'Aiea**. *A rare soft-wooded tree from Hawai'i. There are other species in this genus on other islands.*

34. **Canthium odoratum** *(Rubiaceae, I, 1, T, HI except N & Ka).* **Alahe'e**. *A common tree with fragrant flowers and dark green coffee-like fruits.*

31▲ 32▼

35. **Acacia koaia** *(Fabaceae, Es, 2, T, L, Mo, M, H).* **Koai'a**. *A rare tree with smaller leaves and fruits than the* **koa.**

36. **Diospyros sandwicensis** *(Ebenaceae, Es, 2, T, HI except N, Ka).* **Lama**. *A common tree of lowland dry areas.*

37. **Bidens menziesii** *(Asteraceae, 19, HP, Mo, Mw, H).* **Ko'oko'olau**. *A widespread genus with many Hawaiian species.*

35▲ 36▼                                                         37▶

*38.* **Neraudia melastomifolia** *(Urticaceae, Eg, 7, S, K.).* **Ma'aloa.** *A branched shrub whose fruits turn bright red at maturity.*

*39.* **Delissea rhytidosperma** *(Campanulaceae, Eg, 10, RS, K).* **'Ohā.** *An almost extinct genus of lobelioids more or less restricted to the dry forest.*

*40.* **Euphorbia haeleeleana** *(Euphorbiaceae, Es, 1, T, K, O).* **'Akoko.** *A rather recent discovery from some remote valleys of northwestern Kaua'i and the Wai'anaes on O'ahu.*

38▲ 39▼  40►

41. **Hibiscadelphus distans** (*Malvaceae, Eg, 6, T, K*). **Hau-kuahiwi**. *A rather recent discovery from Waimea Canyon on Kaua'i.*

42. **Streblus pendulinus** (*Moraceae, I, 1, T, HI, except Ka, N*). **A'ia'i**. *A widespread variable species locally common in dry areas.*

43. **Gardenia brighamii** (*Rubiaceae, Es, 3, T, O, Mo, M, L, H*). **Nā'ū**. *A rare shrub, now almost extinct, formerly known from most of the major islands.*

41▲  42▼

43▶

44. **Pittosporum gayanum** (*Pittosporaceae, Es, 11, T, K*). **Hōʻawa**. *A branched shrub from the highlands of Kauaʻi.*

45. **Santalum freycinetianum** (*Santalaceae, Es, 4, S or T, HI except H*). **ʻIliahi**. *Members of this genus are partly parasitic on other plants.*

46. **Hibiscus kokio** (*Malvaceae, Es, 6, T, K, O, Mo, M*). **Kokiʻo** or **kokiʻo-ʻula**. *A rare plant known from Kauaʻi, Oʻahu, Molokaʻi and Maui.*

47. **Alphitonia ponderosa** (*Rhamnaceae, Es, 1, T, HI except N & Ka*). **Kauila**. *A large tree with hard wood, highly valued by the Hawaiians. Rare except on Kauaʻi.*

44▲ 45▼

48. **Isodendrion subsessilifolium** *(Violaceae, Eg, 5, S, K).* **Aupaka**. *A rare genus with many extinct species; the living ones all rare and endangered.*

49. **Nestegis sandwicensis** *(Oleaceae, Es, 1, T, HI except N & Ka).* **Olopua**. *A common tree of the koa forest with purple-black, olive-like fruits.*

50. **Kokia kauaiensis** *(Malvaceae, Eg, 4, T, K).* **Koki'o**. *A rare genus with one extinct species known and the others rare and endangered.*

48▲  49▼                                                                      50▶

86

51. **Remya mauiensis** *(Asteraceae, Eg, 3, S, Mw). A rare genus with all of the species on the verge of extinction.*

52. **Hibiscadelphus hualalaiensis** *(Malvaceae, Eg, 6, T, H).* **Hau-kuahiwi.** *A rare genus with more than half of the species already extinct.*

53. **Psychotria hobdyi** *(Rubiaceae, Es, 11, T, K).* **Kōpiko.** *A rare, attractive small tree from the Kōke'e area of Kaua'i.*

54. **Pisonia sandwicensis** *(Nyctaginaceae, Es, 3, T, HI).* **Pāpala-kēpau.** *Plant with male flowers. A large soft-wooded dioecious tree with sticky fruits.*

51▲ 52▼

55. **Alyxia oliviformis** *(Apocynaceae, Es, 1, V, HI except N and Ka)*. **Maile**. *A common vine of most vegetation zones but most common in mixed mesic forests. It is a very important Hawaiian lei plant.*

56. **Bonamia menziesii** *(Convolvulaceae, Es, 1, V, HI except N & Ka)*. *A rare vining perennial member of the morning-glory family.*

57. **Charpentiera obovata** *(Amaranthaceae, Es, 5, T, HI except Ka & N)*. **Pāpala**. *A soft-wooded tree with drooping panicles of tiny reddish flowers.*

55▲ 56▼                                                          57►

58. **Viola tracheliifolia** (*Violaceae, Es, 8, S, K, O, Mo, M*). **Pāmakani**. *A peculiar woody violet that can be up to 4 m (12 ft) tall.*

59. **Tetraplasandra hawaiensis** (*Araliaceae, Es, 6, T, Mo, L, M, H*). **'Ohe**. *A tall tree with small hairy fruits.*

60. **Claoxylon sandwicense** (*Euphorbiaceae, Es, 1, S, T, HI except N & Ka*). **Po'olā**. *A shrub or small tree of mesic forests.*

61. **Morinda trimera** (*Rubiaceae, Es, 1, T, K, O, L, Mw*). **Noni-kuahiwi**. *An uncommon small tree.*

58▲ 59▼

60▲  61▼

# RAIN FOREST

62. **Clermontia grandiflora** (*Campanulaceae, Eg, 23, S, Mo, M*). **'Ōhā-wai**. *A species characterized by its long, dangling peduncles of green or reddish flowers.*

63. **Clermontia pallida** (*Campanulaceae, Eg, 23, S, Mo*). **'Ōhā-wai**. *A Moloka'i species forming shrubs or small trees on wet, misty ridges.*

64. **Clermontia arborescens** (*Campanulaceae, Eg, 23, S, Mo, M*). **'Ōhā-wai-nui**. *A common species with large, fleshy flowers, sometimes becoming a small tree.*

65. **Clermontia montis-loa** (*Campanulaceae, Eg, 23, S or T, H*). **'Ōhā-wai**. *A common species often hybridizing with* **C**. **parviflora** *in wet forests (following double page).*

62▲ 63▼

64▶

66. **Cyanea bishopii** *(Campanulaceae, Eg, 55, RS, Me)*. **Hā hā**. *A rather small, unbranched plant from the wet forests of East Maui.*

67. **Cyanea superba** *(Campanulaceae, Eg, 55, RS, O)*. **Hā hā**. *A rare species known from only two colonies in the Wai'anaes on O'ahu.*

68. **Cyanea gayana** *(Campanulaceae, Eg, 55, RS, K)*. **Hā hā**. *A large-leaved species which forms a small unbranched shrub on Kaua'i.*

66▲ 67▼                                                                       68▶

69. **Cyanea grimesiana** *(Campanulaceae, Eg, 55, RS, O, Mo, M, L, H)*. **Hā hā**. *A rare species with fern-like foliage.*

70. **Cyanea aculeatiflora** *(Campanulaceae, Eg, 55, RS, Me)*. **Hā hā**. *A prickly species from wet shaded gulches on East Maui.*

71. **Cyanea longipedunculata** *(Campanulaceae, Eg, 55, RS, H)*. **Hā hā**. *A white-flowered species from Hawai'i that is conspicuous in fruit.*

71▶

69▲ 70▼

72. **Lobelia grayana** (*Campanulaceae, Es, 12, RS, Me*). *One of several, closely related blue-flowered species.*

73. **Pelea parvifolia** *var.* **sessilis** (*Rutaceae, Es, 55, S, Mo*). **Alani**. *One of the rarer species of this large genus that is often so dominant in wet forests.*

74. **Hibiscus kahili** (*Malvaceae, Es, 6, S, K*). *A rare and local species from Mt. Kahili, Kaua'i.*

72▲ 73▼ 74►

75. **Joinvillea ascendens** *subsp.* **ascendens**
*(Joinvilleaceae, I, 1, HP, K, O, Mo, M, H).*
**'Ohe**. *A rare reed-like plant whose flowers are
followed by small orange fruits. The subspecies
is endemic but the species is indigenous.*

76. **Wikstroemia furcata** *(Thymelaeaceae, Es,
13, S, K).* **'Ākia**. *One of many Hawaiian species
which are characterized by their tough, fibrous
bark.*

77. **Pritchardia beccariana** *(Arecaceae, Es, 15,
T, H).* **Loulu**. *A tall palm from the wet forests
west of Hilo.*

75▲  76▼                                                                 77▶

78. **Phyllostegia vestita** (*Lamiaceae, Es, 27, HP, H*). *A rare mint from Hawai'i with fleshy greenish-black fruits.*

79. **Hibiscus arnottianus** *var.* **punaluuensis** (*Malvaceae, Es, 6, T, O*). **Koki'o-ke'oke'o.** *A striking shrubby tree from the Ko'olaus on O'ahu.*

80. **Peperomia hesperomanni** (*Piperaceae, Es, 35, HP, K*). **'Ala'alawainui.** *A large genus in Hawai'i with many epiphytic and some terrestial species.*

78▲ 79▼

80▶

81. **Touchardia latifolia** *(Urticaceae, Eg, 1, S, HI except N & Ka).* **Olonā**. *An important source of fiber for the Hawaiians.*

82. **Scaevola glabra** *(Goodeniaceae, Es, 10, S, K, O).* **'Ohe-naupaka**. *An attractive shrub with dark purple fruits restricted to the wet forests of Kaua'i and O'ahu.*

83. **Scaevola mollis** *(Goodeniaceae, Es, 10, S, O).* **Naupaka-kuahiwi**. *A common species of wet ridges and gulches on O'ahu with hairy leaves and white or purplish flowers.*

81▲  82▼

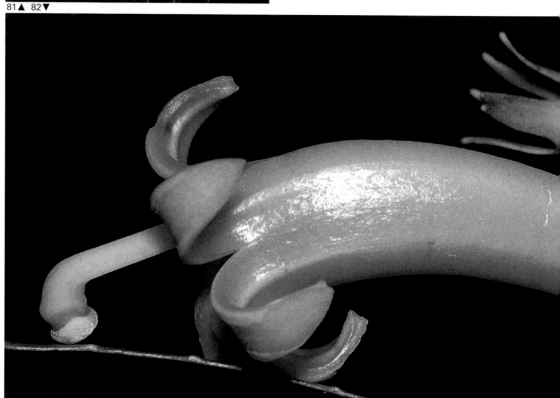

83 ▲

84▲

84. **Platydesma spathulata** (*Rutaceae, Eg, 4, RS, K, O, M, H and probably Mo also*). **Pilo-kea**. *A cauliferous shrub or small tree whose fruits closely resemble those of* **Pelea**.

85. **Pelea anisata** (*Rutaceae, ES, 55, T, K*). **Mokihana**. *An almost vine-like, small shrubby tree with anise-scented fruits prized by the Hawaiians for making leis. It is the lei of Kaua'i Island.*

86. **Hillebrandia sandwicensis** (*Begoniaceae, Eg, 1, HP, K, Mo, Me*). **Aka'aka'awa**. *A rather local plant with a preference for wet, open or shady gulches. It was reported by Hillebrand from Mt. Ka'ala, O'ahu.*

87. **Metrosideros polymorpha** (*Myrtaceae, Es, 7, T, HI except N & Ka*). **'Ohi'a-lehua**. *The dominant tree of the wet forest. The flowers are mostly red, but can occasionally be yellow, and rarely even white.*

88. **Chamaesyce rockii** (*Euphorbiaceae, Es, 14, S, T, O*). **'Akoko**. *An uncommon shrub to small tree with large, bright pink or red capsular fruits (following double page).*

85 ▲  86▼

87▶

89. **Astelia menziesiana** (*Liliaceae, Es, 3, HP, HI except N & Ka*). **Pa'iniu.** *Common epiphytes of the wet forest. Other species exist on other islands. This is a female plant.*

90. **Astelia menziesiana** (*Liliaceae, Es, 3, HP, HI except N & Ka*). **Pa'iniu.** *A purple-flowered form from Moloka'i. This is a male plant.*

91. **Rubus hawaiiensis** (*Rosaceae, Es, 2, S, K, Mo, M, H*). **'Akala.** *A shrub with almost thornless arching branches and with red or yellow fruits.*

92. **Korthalsella complanata** (*Viscaceae, I, 6, P, HI except N & Ka*). **Hulumoa.** *A common parasite on* **koa**, **'Ōhi'a-lehua**, *etc. Native to Henderson Island in the southeast Pacific.*

89▲ 90▼

91▼ 92▲

115

93. **Hesperomannia lydgatei** *(Asteraceae, Eg, 3, T, K). A very rare, almost extinct genus. This species is from southeastern Kaua'i.*

94. **Antidesma platyphyllum** *(Euphorbiaceae, Es, 2, T, HI except N & Ka)* **Hame**. *A common component of the wet forest whose fruits turn red to dark purple at maturity.*

95. **Cyrtandra procera** *(Gesneriaceae, Es, 51, S, Mo).* **Ha'iwale**. *A species found along the margin of Pēpē'ōpae Bog.*

93▲ 94▼                                                                 95►

96. **Cyrtandra calpidicarpa** *(Gesneriaceae, Es, 51, S, O).* **Ha'iwale**. *One of many species occuring in shady gulches in the Ko'olaus on O'ahu.*

97. **Cyrtandra platyphylla** *(Gesneriaceae, Es, 51, S, H).* **Ha'iwale**. *A common understory plant of Maui and Hawai'i.*

98. **Psychotria mariniana** *(Rubiaceae, Es, 11, S or T, K, O, Mo, L, M).* **Kōpiko**. *A common plant of mesic to wet forest areas.*

99. **Gouldia affinis** *(Rubiaceae, Eg, 4, S, HI).* **Manono**. *A highly variable, common wet forest plant. The fruits are purple-black.*

100. **Ilex anomala** *(Aquifoliaceae, I, 1, S or T, HI except Ka & N).* **Kāwa'u**. *A common shrub or small tree of mesic to wet forests, and occasionally in bogs.*

96▲ 97▼

101. **Bobea elatior** (*Rubiaceae, Eg, 3, T, O, Mo, L, M, H*). **'Ahakea**. *A wet forest tree recognized by its pale green foliage.*

102. **Labordia hedyosmifolia** (*Loganiaceae, Eg, 7, S, O, M, H*). **Kāmakahala**. *A small shrub with odd-shaped capsular fruits.*

103. **Dubautia knudsenii** (*Asteraceae, Eg, 21, T, K*). **Na'ena'e**. *A rare, small tree from northwestern Kaua'i.*

104. **Freycinetia arborea** (*Pandanaceae, Es, 1, V, HI except N & Ka*). **'Ie'ie**. *A common woody climber in wet areas.*

101▲  102▼

**105. Hedyotis fluviatilis** *var.* **kamapuuana** *(Rubiaceae, Es, 20, S, O).* **Kamapua'a**. *A rather rare scandent shrub with foetid foliage & flowers.*

**106. Pipturus albidus** *(Urticaceae, Es, 1, S or T, HI except N & Ka).* **Māmaki**. *A common plant of wet mesic areas throughout the islands. It is one of the sources of bark for tapa.*

**107. Broussaisia arguta** *(Hydrangeaceae, Eg 1, S, HI except N & Ka).* **Kanawao**. *A common shrub of wet areas. The male and female flowers are on separate plants. Female shown.(following double page).*

105▼  106▶

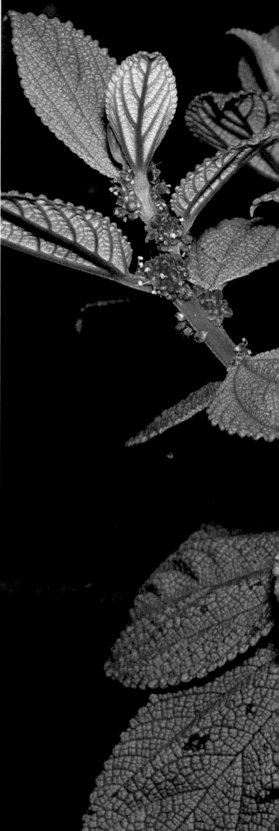

# BOGS

108. **Argyroxiphium grayanum** (*Asteraceae, Eg, 5, RS, M*). *A greensword that is often branched, and appearing shrubby. It is found at the margins of bogs, in open bogs and in rain forests.*

109. **Argyroxiphium caliginis** (*Asteraceae, Eg, 5, RS, Mw*). **'Āhinahina**. *Unlike the Haleakalā silversword, this reproduces vegetatively by series of offshoots at the base of the older plants.*

110. **Argyroxiphium caliginis**. **'Āhinahina**. *Close-up of the inflorescence.*

108▲ 109▼

110▶

111. **Viola kauaensis** *(Violaceae, Es, 8, HP, K, O).* **Nani-wai'ale'ale**. *One of two trailing bog species found in the islands.*

112. **Drosera anglica** *(Droseraceae, I, 1, HP, K).* **Mikinalo**. *Hawai'i's only insectivorous plant is confined to the bogs and swampy areas of Kaua'i.*

113. **Lycopodium cernuum** *(Lycopodiaceae, I, 9, HP, HI).* **Wāwae'iole**. *These plants creep over the surface of open bogs but in forests they grow upright. In the photo are also* **Rhynchospora lavarum (kuolohia)** *and* **Oreobolus furcatus.**

114. **Lobelia kauaensis** *(Campanulaceae, Es, 12, RS, K).* **Pu'e**. *Inflorescences 3–4 ft tall. Other species are found on O'ahu, Moloka'i, and Maui.*

111▲

112 ▲ 113▼

114▶

115. **Dubautia paleata** *(Asteraceae, Eg, 21, S, K).* **Na'ena'e-puakea.** *Grows in and at the margins of boggy areas of the Alaka'i Swamp.*

116. **Stenogyne kamehamehae** *(Lamiaceae, Eg, 20, HP, Mo, Mw). Occasionally trailing over swampy ground near the margins of bogs, but more commonly found in wet forest.*

117. **Oreobolus furcatus** *(Cyperaceae, Es, 1, HP, K, M, Mo, H). A tussock-forming sedge found in boggy areas on all of the major islands.*

118. **Dichanthelium** *(Panicum)* **cynodon** *(Poaceae, 5, HP, Mo). A tussock-forming grass from Moloka'i's Pēpē'ōpae Bog. It is surrounded by miniature* **Metrosideros** *plants.*

115▲ 116▼

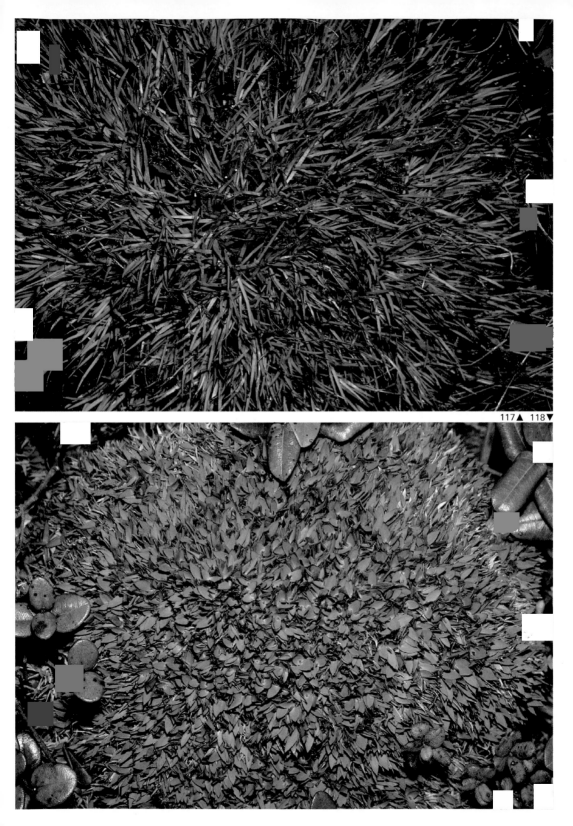

119. **Chamaesyce remyi** *(Euphorbiaceae, Es, 14, S, K).* **'Akoko**. *A variable plant, this particular one is from the Wahiawa Bog on Kaua'i.*

120. **Plantago melanochrous** *(Plantaginaceae, Es, 2–9, HP, Mw).* **Laukāhi-kuahiwi**. *The bog plantagos are taxonomically confused and in need of study.*

121. **Lobelia gloria-montis** *(Campanulaceae, Es, 12, RS, Mo, M). A striking member of the lobelia family.*

119▲ 120▼

121▶

# ALPINE PLANTS

122▲

123▲  124▼

122. **Tetramolopium humile** (Asteraceae, Es, 11, S, Me, H). **Pāmakani**. *Low, decumbent perennial from Mauna Kea and Haleakalā.*

123. **Dubautia menziesii** (Asteraceae, Eg, 21, S, Me). **Na'ena'e**. *Low shrub with succulent leaves and orange-yellow flowers. It is common at the summit of Haleakalā.*

124. **Gnaphalium sandwicensium** *var.* **kilaueanum** (Asteraceae, Es, 1, HP, H). **'Ena'ena**. *Frequent along the Saddle Road and the vicinity of Kīlauea on Hawai'i. It also occurs in the Ka'u Desert. Other varieties occur on other islands.*

125. **Argyroxiphium kauense** (Asteraceae, Eg, 5, RS, H). **'Āhinahina**. *This species differs from the Haleakalā and Mauna Kea silverswords in the caulescent habit and thinner, less silvery leaves. It is a rare plant found on Mauna Loa.*

126. **Argyroxiphium sandwicense** *subsp.* **macrocephalum** (Asteraceae, Eg, 5, RS, Me). **'Āhinahina**. *Differs from the Mauna Kea plants in the shorter and stouter inflorescence and larger flower heads.*

125▼  126▶

127. **Silene hawaiiensis** (*Caryophyllaceae, Es, 6, S, H*). *Low, sprawling shrub branching from a parsnip-like root. The flowers are nocturnal.*

128. **Silene hawaiiensis**. *Close-up of flowers.*

129. **Sophora chrysophylla** (*Fabaceae, Es, 1, S or T, HI except N, Ka*). **Māmane**. *A dominant shrub or small tree of the alpine areas, but it is also found in mixed mesic forest.*

130. **Vaccinium reticulatum** (*Ericaceae, Es, 3, S, M, H*). **'Ōhelo**. *This is a common plant with berries that are pale orange to dark red in color (following double page).*

127▲ 128▼

129►

131. **Sisyrinchium acre** *(Iridaceae, Es, 1, HP, M, H).* **Mau'ulā'ili**. *Formerly used by the Hawaiians in tatooing.*

132. **Santalum haleakalae** *(Santalaceae, Es, 4, T, P, Me).* **'Iliahi**. *Dense small tree with flowers turning red at maturity followed by purple-black fruits.*

131▼  132▶

133. **Stenogyne microphylla** *(Lamiaceae, Eg, 20, S, H). A scandent, diffusely branched shrub from the lower alpine zone on Hawai'i.*

134. **Coprosma ernodeoides** *(Rubiaceae, Es, 15, S, Me, H).* **Kūkaenenē**. *This is a staminate or male plant. It is a trailing or scandent shrub on lava with shiny black fruits.*

135. **Geranium cuneatum** *subsp.* **hololeucum** *(Geraniaceae, Es, 5, S, H).* **Nohoanu** *or* **hinahina**. *Another subspecies occurs on Haleakalā with even more silvery leaves.*

136. **Geranium arboreum** *(Geraniaceae, Es, 5, S, Me).* **Nohoanu** *or* **hinahina**. *A rare shrub which never becomes tree-like. It is from the lower alpine zone of Haleakalā.*

133▲  134▼

143

# PLANTS OF LAVA FLOWS

137. **Sadleria cyatheoides** *(Blechnaceae, I, 6, HP, HI except Ka & N).* **Ama'u**. *A colonizer of young lava flows.*

138. **Stereocaulon** *sp.* **Limu-haea**. *A related species colonizes new lava flows simultaneously with flowering plants.*

139. **Dubautia scabra** *(Asteraceae, Eg, 21, HP, Mo, L, M, H).* **Na'ena'e** *or* **kūpaoa**. *A low decumbent plant with white flowers easily seen on recent flows on Kīlauea.*

140. **Polypodium pellucidum** *(Polypodiaceae, Es, 1, HP, HI except N, Ka).* **'Ae**. *A tough, leathery fern which is common at Kīlauea.*

137▲ 138▼

139▼ 140▶

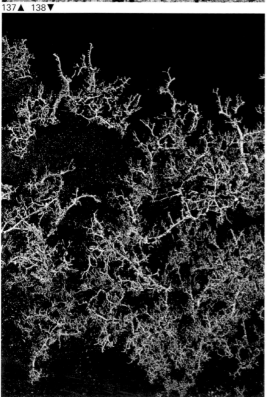

# INTRODUCED PLANTS

*141.* **Oxalis corniculata** *(Oxalidaceae). A common roadside, street and garden weed.*

*142.* **Tournefortia argentea** *(Boraginaceae). A common exotic plant of the coastal zone.*

*143.* **Passiflora mollissima** *(Passifloraceae). A pernicious vine rapidly spreading throughout forests on Kaua'i and Hawai'i.*

*144. Guava* **(Psidium guajava)***. Kuawa. Widespread in Hawai'i at lower elevations in particular where it often forms dense thickets (following double page).*

141▲ 142▼

143▶

145 and 146. **Koa-haole (Leucaena leucocephala**). *A roadside shrub common in lowlands and lower mountain slopes throughout Hawai'i, often to the exclusion of anything else.*
*a) View of plant.*
*b) A shrubby jungle formed by the species.*

147. **Lantana camara** *(Verbenaceae). A shrubby plant that has taken very well to drier habitats of Hawai'i.*

148. **Aleurites moluccana** *(Euphorbiaceae)* **kukui.** *Its light-colored foliage is conspicuous in Hawaiian forests.*

145▲  146▼

149. **Kalanchöe pinnata** (*Crassulaceae*). Another plant escaped from cultivation in Hawai'i. Commonly called the air plant.

150. *Kahili ginger* (**Hedychium gardnerianum**). *One of three or four species of ginger that have extensively naturalized in Hawaiian forests.*

151. **Cocos nucifera; niu**. *The coconut palm was brought to the islands by the Polynesians (following page).*

149▼  150▶

# INDEX TO SCIENTIFIC NAMES

156

# BIBLIOGRAPHY

## LITERATURE CITED

ARMSTRONG, R. W. 1973. Atlas of Hawaii. *Univ. Press Hawaii, Honolulu, 272 pp.*
CARLQUIST, S. 1970. Hawaii, a Natural History. *Natural History Press, New York, 463 pp.*
CARLQUIST, S. 1974. Island biology. *Columbia Univ. Press, New York, 660 pp.*
FOSBERG, F. R. 1948. Derivation of the flora of the Hawaiian Islands: *in Zimmerman, E. C., Insects of Hawaii. Vol. 1. Univ. Press of Hawaii, Honolulu, pp. 107-119.*
KAY, E. A. (ed.). 1972. A Natural History of the Hawaiian Islands. Selected readings. *Univ. Press Hawaii, Honolulu, 653 pp.*
OLSON, S. L., and H. F. JAMES. 1984. The role of Polynesians in the extinction of the avifauna of the Hawaiian Islands: *in Martin, P. S., and R. G. Klein (eds.), Quaternary extinctions, a prehistoric revolution. Univ. Arizona Press, Tucson.*
RIPPERTON, J. C., and E. Y. HOSAKA. 1942. Vegetation zones of Hawai'i. *Hawaii Agric. Exp. Sta. Bull. 89: 1-60.*
ROCK, J. F. 1913. The Indigenous Trees of the Hawaiian Islands. *Publ. privately, 518 pp.* (Reprinted, with introduction and addenda, Chas. E. Tuttle Co., Rutland, Vt., 548 pp., 1974).
ROCK, J. F. 1919. A monographic study of the Hawaiian species of the tribe Lobelioideae, family Campanulaceae. *Mem. Bernice P. Bishop Mus. 7: 1-395.*
WAGNER, W. L., and W. GAGNE. A summary of the derivation, speciation, and endemism of the Hawaiian biota. *Hawaii Botanical Society Newsletter* (in press).

## GENERAL REFERENCES TO THE HAWAIIAN FLORA

BALGOOY, M. M. J. van (ed.). 1975. Pacific plant areas, Vol. 3. *Rijksherbarium, Leiden, 386 pp.*
BALGOOY, M. M. J. van (ed.). 1984. Pacific plant areas, Vol. 4. *Rijksherbarium, Leiden, 270 pp.*
CARLQUIST, S. 1980. Hawaii, a Natural History, ed. 2. Geology, climate, native flora and fauna above the shoreline. *Pacific Trop. Bot. Gard, Honolulu, 468 pp.*
DEGENER, O. 1932–1980. Flora Hawaiiensis or New Illustrated Flora of the Hawaiian Islands. *Publ. privately, Honolulu.* [See Mill et al., Taxon 34: 229-259 (1985) for citation of each of the 1,144 articles included in this work].
DEGENER, O. 1945. Plants of Hawaii National Park illustrative of plants and customs of the South Seas. [Revised ed. of Illustrated guide to the more common or noteworthy ferns and flowering plants of Hawaii National Park, 1930]. *Publ. privately, Honolulu, 314 pp.*
FOSBERG, F. R., and D. HERBST. 1975. Rare and endangered species of Hawaiian vascular plants. *Allertonia 1: 1-72.*
HILLEBRAND, W. 1888. Flora of the Hawaiian Islands. *Lubrecht & Cramer, Monticello, NY, 673 pp.* (Facsimile ed., B. Westermann & Co., New York, 1981).
MUELLER-DOMBOIS, D., K. W. BRIDGES, and H. L. CARSON. 1981. Island ecosystems. Biological organization in selected Hawaiian communities. *US/IBP Synthesis Series 15. Hutchinson Ross Publ. Co., Woods Hole, Mass., 583 pp.*
NEAL, M. C. 1965. In gardens of Hawaii, ed. 2. *Special Publ. Bernice P. Bishop Mus. 50: 1-924.*
PORTER, J. R. 1972. Hawaiian names for vascular plants. *Department paper 1, College of Tropical Agriculture, University of Hawaii.*
ST. JOHN, H. 1973. List and summary of the flowering plants in the Hawaiian Islands. *Pacific Trop. Bot. Gard. Mem. 1: 1-519.*
STEENIS, C. G. G. J. van (ed.). 1963. Pacific plant areas, Vol. 1. *Monogr. Natl. Inst. Sci. Technol. (Manila) 8, Vol. 1: 1-297.*
STEENIS, C. G. G. J. van, and M. M. J. van BALGOOY (eds.). 1966. Pacific plant areas, Vol. 2. *Blumea Suppl. 5: 1-312.*
STONE, B. C. 1967. A review of the endemic genera of Hawaiian plants. *Bot. Rev. (Lancaster) 33: 216-259.*
STONE, C. P., and J. M. SCOTT (eds.). 1985. Hawai'i's terrestrial ecosystems: preservation and management. *Coop. Natl. Park Resources Stud. Unit, Univ. Hawaii, 584 pp.*
WAGNER, W. L., D. HERBST, and S. H. SOHMER, Manual of the Flowering Plants of Hawai'i (in press).

159

# ACKNOWLEDGEMENTS

The information presented in this book, particularly that concerning the identity, names, abundance and distribution of the species mentioned, as well as the overall Hawaiian vegetation patterns, would not have been anywhere near as accurate and consistent as it probably is were it not for Drs. Derral Herbst and Warren Wagner. These two, who are the principal authors of the Bishop Museum's project to produce a *Manual of the Flowering Plants of Hawai'i*, provided a great deal of support for this endeavour. Dr. Isabella Abbott provided similar support for the ethnobotanical portion, and Dr. Gerald Carr provided information concerning hybridization in the Hawaiian Silversword complex. Michael McKenney, who has one of the best grasps of Hawaiian names for native Hawaiian plants at present supplied the Hawaiian names used here. In addition we wish to thank Christa Russell, Ken and Ron Nagata, Lee Motteler, Drs. Yosi Sinoto and JoAnn Tenorio, and Rylan Yee who contributed time or answered questions. We acknowledge also the typing efforts of Shirley Rosenbush, and the typing and editorial help provided by Susan Mill. We also wish to acknowledge the support provided by the Natural History Museum, Los Angeles County.

## PHOTO CREDITS

*Bishop Museum* pp. 6 (map), 10, 11.
*Gerald Carr* pp. 32, 35, 37 (top)
*Wayne Gagne* pp. 13, 37 (bottom)
*Frank Holmes Laboratories* p. 9
*Helen Kennedy* p. 24
*Ken Nagata* p. 34
*S. H. Sohmer* pp. 19, 31, 40
*Warren L. Wagner* pp. 117, 118 (No. 97), 146, 150, 151 (No. 147), 152
*Laurel Woodley* pp. 64 (No. 10), 75 (No. 30)
*Robert Gustafson*    All other photos except those credited above

### Scientific Editor: Bernard SALVAT